THE CHEMISTRY OF BEER

The Science in the Suds

THE CHEMISTRY OF BEER
The Science in the Suds

ROGER BARTH, PhD

Photography by Marcy Barth.

Copyright © 2013 by John Wiley & Sons, Inc. All rights reserved

Published by John Wiley & Sons, Inc., Hoboken, New Jersey
Published simultaneously in Canada

No part of this publication may be reproduced, stored in a retrieval system, or transmitted in any form or by any means, electronic, mechanical, photocopying, recording, scanning, or otherwise, except as permitted under Section 107 or 108 of the 1976 United States Copyright Act, without either the prior written permission of the Publisher, or authorization through payment of the appropriate per-copy fee to the Copyright Clearance Center, Inc., 222 Rosewood Drive, Danvers, MA 01923, (978) 750-8400, fax (978) 750-4470, or on the web at www.copyright.com. Requests to the Publisher for permission should be addressed to the Permissions Department, John Wiley & Sons, Inc., 111 River Street, Hoboken, NJ 07030, (201) 748-6011, fax (201) 748-6008, or online at http://www.wiley.com/go/permissions.

Limit of Liability/Disclaimer of Warranty: While the publisher and author have used their best efforts in preparing this book, they make no representations or warranties with respect to the accuracy or completeness of the contents of this book and specifically disclaim any implied warranties of merchantability or fitness for a particular purpose. No warranty may be created or extended by sales representatives or written sales materials. The advice and strategies contained herein may not be suitable for your situation. You should consult with a professional where appropriate. Neither the publisher nor author shall be liable for any loss of profit or any other commercial damages, including but not limited to special, incidental, consequential, or other damages.

For general information on our other products and services or for technical support, please contact our Customer Care Department within the United States at (800) 762-2974, outside the United States at (317) 572-3993 or fax (317) 572-4002.

Wiley also publishes its books in a variety of electronic formats. Some content that appears in print may not be available in electronic formats. For more information about Wiley products, visit our web site at www.wiley.com.

Library of Congress Cataloging-in-Publication Data:

Barth, Roger.
  The chemistry of beer: the science in the suds / Roger Barth, Ph.D.
      pages cm
  Includes index.
  ISBN 978-1-118-67497-0 (paper)
  1. Beer. 2. Beer–Analysis. I. Title.
  TP577.B35 2013
  663'.42–dc23
                          2013013982

Printed in the United States of America

20 19 18 17 16 15 14 13 12

CONTENTS

PREFACE		xi
ACKNOWLEDGMENTS		xiii
ABOUT THE AUTHOR		xvii
PERIODIC TABLE OF ELEMENTS		xviii
1	**INTRODUCTION**	**1**
	1.1 Brief History	1
	1.2 The World of Beer	7
	1.3 Beer and Chemistry	9
	1.4 Alcohol and Prohibition	10
	1.5 Beer Tradition	12
	Bibliography	13
	Questions	15
2	**WHAT IS BEER?**	**17**
	2.1 Beer Ingredients	17
	2.2 Beer as Food	21
	2.3 How Beer Is Made	23
	Bibliography	31
	Questions	31

3 CHEMISTRY BASICS — 33

- 3.1 Atoms — 33
- 3.2 Energy Levels and the Periodic Table — 34
- 3.3 Compounds — 36
- 3.4 Ionic Bonds — 38
- 3.5 Covalent Bonds and Molecules — 40
- 3.6 Molecular Shape — 43
- 3.7 Polarity and Electronegativity — 46
- 3.8 Intermolecular Forces — 48
- 3.9 Molecular Kinetics — 51
- 3.10 Chemical Reactions and Equations — 52
- 3.11 Mixtures — 53
- Bibliography — 53
- Questions — 53

Appendix to Chapter 3: Measurement in Chemistry — 56

- Numbers — 56
- International System — 57
- Mass Relationships in Compounds — 63
- Composition of Mixtures — 64
- Bibliography — 66
- Questions — 67

4 WATER — 69

- 4.1 The Water Molecule — 69
- 4.2 Acids and Bases — 71
- 4.3 pH — 73
- 4.4 Ions and Beer — 76
- 4.5 Water Treatment — 79
- Bibliography — 85
- Questions — 86

5 INTRODUCTION TO ORGANIC CHEMISTRY — 89

- 5.1 Structural Formulas — 89
- 5.2 Functional Groups — 91
- 5.3 Using the Functional Group Guide — 107
- Bibliography — 110
- Questions — 110

6 SUGARS AND STARCHES — 113

- 6.1 Monosaccharides — 113
- 6.2 Chirality — 115
- 6.3 Disaccharides — 120
- 6.4 Polysaccharides — 121
- Bibliography — 126
- Questions — 127

7 MILLING AND MASHING — 131

- 7.1 Milling — 132
- 7.2 Mashing — 133
- 7.3 Enzymes and Proteins — 135
- 7.4 Mashing Process — 141
- 7.5 Dextrins, Light Beer, and Malt Liquor — 146
- Bibliography — 146
- Questions — 147

8 WORT SEPARATION AND BOILING — 149

- 8.1 Wort Separation — 149
- 8.2 Boiling — 152
- 8.3 Hops — 153
- 8.4 Hot Break — 155
- 8.5 Chilling — 158
- Bibliography — 159
- Questions — 159

9 FERMENTATION — 161

- 9.1 The Anatomy of Brewing — 161
- 9.2 Energy and Bonds — 165
- 9.3 Glycolysis — 168
- 9.4 Ethanol Synthesis — 170
- 9.5 Aerobic and Anaerobic Reactions — 170
- 9.6 Higher Alcohols — 172
- 9.7 Esters — 173
- Bibliography — 174
- Questions — 175

10 TESTS AND MEASUREMENTS — 177

- 10.1 Carbohydrate Content — 177
- 10.2 Temperature — 183
- 10.3 Color — 185
- 10.4 Alcohol Content — 187
- 10.5 pH — 190
- 10.6 Sensory Analysis — 192
- Bibliography — 192
- Questions — 192

11 THE CHEMISTRY OF FLAVOR — 195

- 11.1 Anatomy of Flavor — 195
- 11.2 Taste — 196
- 11.3 Aroma — 198
- 11.4 Mouth Feel — 200
- 11.5 Flavor Units — 201
- 11.6 Flavor Compounds in Beer — 202
- Bibliography — 207
- Questions — 208

12 THE CHEMISTRY OF BEER STYLES — 211

- 12.1 Beer Style Families — 211
- 12.2 Realizing a Style — 215
- Bibliography — 222
- Questions — 223

13 FOAM AND HAZE — 225

- 13.1 Surfaces — 225
- 13.2 Surface Energy — 225
- 13.3 Surfactants — 227
- 13.4 Haze — 227
- 13.5 Foam — 231
- 13.6 Foam Issues — 236
- 13.7 Nitrogen and Widgets — 237
- Bibliography — 238
- Questions — 238

14	**BEER PACKAGING**		**241**
	14.1	Casks and Kegs	241
	14.2	Glass	243
	14.3	Metals	244
	14.4	Aluminum	245
	14.5	Bottling and Canning	247
	14.6	Microbe Reduction	249
		Bibliography	249
		Questions	249
15	**BEER FLAVOR STABILITY**		**251**
	15.1	Typical Flavor Changes	251
	15.2	The Role of Oxygen	252
	15.3	Staling Prevention	260
		Bibliography	262
		Questions	263
16	**BREWING AT HOME**		**265**
	16.1	Safety Issues	266
	16.2	Full Mash Brewing	267
	16.3	Full Mash Brewing Procedure	277
	16.4	Extract Brewing	283
	16.5	Bottling	286
	16.6	Starter Brewing Systems	288
	16.7	Recipes	290
		Bibliography	293
		Questions	293
GLOSSARY			**295**
INDEX			**317**

PREFACE

Whether you are a serious brewer or a person who is just interested in beer, the more you know about the scientific basis of beer, its preparation, and its flavors, the more you will appreciate and enjoy the depth and diversity of the world of beer. Although this book is written largely for the general reader, there is material that will be useful to brewers or people who are considering taking up brewing as homebrewers or as professionals. What distinguishes this book from others on the topic is the logic and sequence of the presentation of chemistry concepts, first atoms, then electrons, then chemical bonds, then molecules.

"Measurement in Chemistry," the Appendix to Chapter 3, includes units of measure, the mole concept, and mass calculations from chemical formulas. Water alkalinity and hardness measurements are dealt with in Chapter 4. Specific gravity tables and hydrometer corrections are presented in Chapter 10. The basics of computing a beer recipe are covered in Chapter 12. Chapter 16 includes some basic homebrewing recipes. Each chapter has references to some of the key primary and secondary literature and questions intended to help you study the material. Questions marked with an asterisk (*) are more challenging and may depend on supplemental material. Although commercial brewing and homebrewing are discussed to help give context to the material, this is not intended to be a complete textbook on brewing, several of which are mentioned in the chapter bibliographies. Nonetheless, it will be very helpful to read this book before one reads one of the brewing textbooks.

Many readers will be college students, some of whom are in the process of developing attitudes and practices regarding the use of alcohol. Alcohol is what it is. It has enriched many lives and ruined many lives, making it much like every other aspect of the human experience. It should not be taken lightly. I hope this book will help its readers attain a thoughtful approach to alcohol. Those who are experiencing difficulties with alcohol should seek assistance from college, pastoral, or health counselors.

Roger Barth, Ph.D.

Department of Chemistry
West Chester University
West Chester, Pennsylvania

ACKNOWLEDGMENTS

Many persons provided support and assistance that greatly enhanced the quality of this book. I deeply appreciate their irreplaceable contributions. Of those, I can acknowledge only a few. Justin Ligi of the West Chester University English Department put in many hours of effort assuring that the chemistry could be understood. Mario and Donna Marie Zoccoli, my eagle-eyed copy readers, went over the manuscript multiple times correcting typographical and stylistic errors and making suggestions for improved readability. Fellow chemists Douglas Hauser (Rutgers), David Cichowicz (La Salle), and Joel Ressner (West Chester) made helpful corrections and suggestions to the chemistry of early versions of the manuscript. Joe Frinzi and his staff at the Mill Creek Avenue Brewery of Yuengling Beer Company in Pottsville, Pennsylvania, gave us a detailed tour. Our local experimental malting barley farmers, Bryan Taylor and Matthew Canan, provided insights as well as samples. David Wilson of Alaskan Brewing Company provided insights from his experience with mash filtration. The librarians at West Chester University, particularly Walter Cressler, did outstanding work tracking down references. Anthropologist Diane Freedman of Philadelphia Community College provided many corrections and helpful suggestions for the first chapter. Regrettably, she passed away a few days before the manuscript was submitted. Gary Beauchamp of the Monel Chemical Senses Center provided helpful insights for Chapter 11. Whitney Thompson of Victory Brewing Company in Downingtown, Pennsylvania, conducted us through the brewery and kindly allowed Marcy to take splendid photographs. Larry Horwitz of the Iron Hill Brewery in West Chester, Pennsylvania, got me interested in flavor chemistry. The Carlsberg Brewery in Ashkelon, Israel, gave me special access to their unique museum/visitor center. Anat Meir, Carlsberg's Laboratory Manager, showed me the brew house and laboratory of this ultramodern facility. Fellow author Patrick McGovern (*Uncorking the Past*) provided valuable assistance for the first chapter, and Don Russell (*Joe Sixpack* column in the Philadelphia Daily News) suggested the subtitle of this book.

The visual appeal of this book is due to the artistry and skilled work of my talented wife, Marcy Barth. All photographs not credited to a specific source are hers. Her constant support made this project possible.

R. B.

ABOUT THE AUTHOR

Roger Barth was born in New York City. He attended public schools in Levittown, Pennsylvania, and received his bachelor's degree from La Salle College in Philadelphia. He was awarded a doctorate in physical chemistry at the Johns Hopkins University. After working at UOP Inc. in Des Plaines, Illinois, he did postdoctoral work at University of Delaware and at Drexel University. He has been teaching chemistry at West Chester University of Pennsylvania since 1985, and he created the Chemistry of Beer course in 2009.

Periodic Table of the Elements

1A																	8A
1 H Hydrogen 1.008	2A											3A	4A	5A	6A	7A	2 He Helium 4.003
3 Li Lithium 6.941	4 Be Beryllium 9.012											5 B Boron 10.811	6 C Carbon 12.011	7 N Nitrogen 14.007	8 O Oxygen 15.999	9 F Fluorine 18.998	10 Ne Neon 20.180
11 Na Sodium 22.990	12 Mg Magnesium 24.305											13 Al Aluminum 26.982	14 Si Silicon 28.086	15 P Phosphorus 30.974	16 S Sulfur 32.066	17 Cl Chlorine 35.453	18 Ar Argon 39.948
19 K Potassium 39.098	20 Ca Calcium 40.078	21 Sc Scandium 44.956	22 Ti Titanium 47.867	23 V Vanadium 50.942	24 Cr Chromium 51.996	25 Mn Manganese 54.938	26 Fe Iron 55.845	27 Co Cobalt 58.933	28 Ni Nickel 58.693	29 Cu Copper 63.546	30 Zn Zinc 65.38	31 Ga Gallium 69.723	32 Ge Germanium 72.61	33 As Arsenic 74.922	34 Se Selenium 78.96	35 Br Bromine 79.904	36 Kr Krypton 83.80
37 Rb Rubidium 85.468	38 Sr Strontium 87.62	39 Y Yttrium 88.906	40 Zr Zirconium 91.224	41 Nb Niobium 92.906	42 Mo Molybdenum 95.96	43 Tc Technetium [98]	44 Ru Ruthenium 101.07	45 Rh Rhodium 102.906	46 Pd Palladium 106.42	47 Ag Silver 107.868	48 Cd Cadmium 112.411	49 In Indium 114.818	50 Sn Tin 118.710	51 Sb Antimony 121.760	52 Te Tellurium 127.60	53 I Iodine 126.904	54 Xe Xenon 131.29
55 Cs Cesium 132.906	56 Ba Barium 137.327	57 La Lanthanum 138.906	72 Hf Hafnium 178.49	73 Ta Tantalum 180.948	74 W Tungsten 183.84	75 Re Rhenium 186.207	76 Os Osmium 190.2	77 Ir Iridium 192.22	78 Pt Platinum 195.084	79 Au Gold 196.967	80 Hg Mercury 200.59	81 Tl Thallium 204.383	82 Pb Lead 207.2	83 Bi Bismuth 208.980	84 Po Polonium [209]	85 At Astatine [210]	86 Rn Radon [222]
87 Fr Francium [223]	88 Ra Radium [226]	89 Ac Actinium [227]	104 Rf Rutherfordium [265]	105 Db Dubnium [268]	106 Sg Seaborgium [271]	107 Bh Bohrium [270]	108 Hs Hassium [277]	109 Mt Meitnerium [276]	110 Ds Darmstadtium [281]	111 Rg Roentgenium [280]	112 Cn Copernicium [285]						

58 Ce Cerium 140.116	59 Pr Praseodymium 140.908	60 Nd Neodymium 144.242	61 Pm Promethium 144.91	62 Sm Samarium 150.36	63 Eu Europium 151.965	64 Gd Gadolinium 157.25	65 Tb Terbium 158.925	66 Dy Dysprosium 162.50	67 Ho Holmium 164.930	68 Er Erbium 167.26	69 Tm Thulium 168.934	70 Yb Ytterbium 173.054	71 Lu Lutetium 174.967
90 Th Thorium 232.038	91 Pa Protactinium 231.036	92 U Uranium 238.029	93 Np Neptunium [237]	94 Pu Plutonium [244]	95 Am Americium [243]	96 Cm Curium [247]	97 Bk Berkelium [247]	98 Cf Californium [251]	99 Es Einsteinium [252]	100 Fm Fermium [257]	101 Md Mendelevium [258]	102 No Nobelium [259]	103 Lr Lawrencium [262]

CHAPTER 1

INTRODUCTION

Beer! This foamy, refreshing, sparkling alcoholic beverage conjures images of parties, festivals, sporting events, and generally fun stuff. Beer is as much a symbol of our culture as football or ballet.

1.1 BRIEF HISTORY

Beer Origins

The origins of **beer** go back to the origins of civilization. Excavations at a prehistoric town, called **Godin Tepe**, located on the ancient Silk Road in the Zagros Mountains (Fig. 1.1) in what is now western Iran uncovered a 5500 year old pottery jar containing calcium oxalate (CaC_2O_4). Calcium oxalate is the signature of beer production. Although there is earlier evidence of mixed fermented beverages, the find at Godin Tepe is the earliest chemical evidence for the brewing of barley beer.

The history of beer is as old as history itself. History begins in **Sumer** (SOO mer), a civilization of city states in southeastern **Mesopotamia** (now Iraq) at the downstream end of the Tigris and Euphrates Rivers. Sumer is the site of the first known written language. Sumerian and other Mesopotamian languages were written with symbols, called **cuneiform**, made with a wedge-shaped stylus often pressed into moist clay tablets. Clay is durable; many ancient cuneiform documents survive and have been translated. Among the

---

The Chemistry of Beer: The Science in the Suds, First Edition. Roger Barth.
© 2013 John Wiley & Sons, Inc. Published 2013 by John Wiley & Sons, Inc.

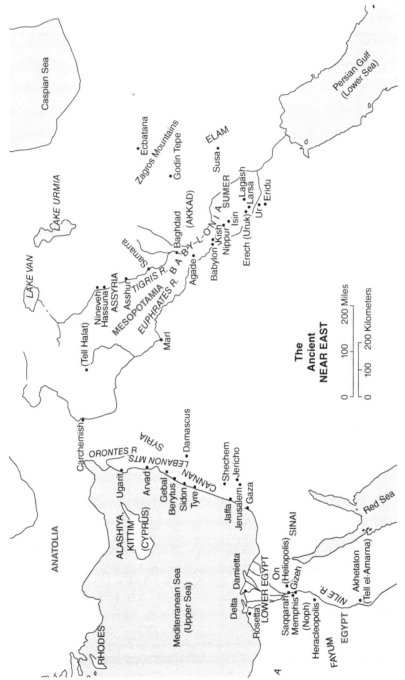

Figure 1.1 Map. *Source:* Adapted from CIA World Factbook. (See color insert.)

Figure 1.2 A 2600 year old Sumerian–Akkadian dictionary of brewing terms, Metropolitan Museum of Art, New York. (See color insert.)

earliest of these are documents written 5000 years ago concerning the brewing and consumption of beer (Fig. 1.2). These tablets record an already mature brewing culture, showing that beer was old when writing was new. One famous cuneiform tablet from 4000 years ago has a poem called *Hymn to Ninkasi*, a hymn of praise to the Sumerian goddess of beer. The *Hymn* has a poetic but not completely comprehensible account of how beer was made. The Sumerians made beer from bread and **malt** (or maybe malt bread), and flavored it, perhaps with honey. Sumerian documents mention beer frequently, especially in the context of temple supplies. Beer was also considered a suitable vehicle for administering medicinal herbs.

Babylon and Egypt

Dominance over the Mesopotamian region passed back and forth among the Sumerian cities until they were conquered by Hammurabi of **Babylon** 3700 years ago. Babylon was a city on the Euphrates upriver from Sumer. Beer made from barley or emmer (an ancient form of wheat) was a staple of the Babylonian diet. After Babylon came Assyria, ruling the Middle East from its two capitals of Asshur and Nineveh. The Assyrians were displaced by a

Figure 1.3 Ancient Egyptian tile: grinding grain to packaging beer, Carlsberg–Israel Visitor Center, Ashkelon, Israel. (See color insert.)

second wave of Babylonians, among whom was Nebuchadnezzar of Biblical infamy. These cultures continued the brewing tradition of Sumer. It is believed that brewing spread from the Mesopotamian region to Egypt, about 800 miles (1300 kilometers) away in Africa. Beer was the primary beverage in Egypt at all levels from the Pharaoh to the peasants. The dead were buried with supplies of beer. Mourners of deceased nobles brought offerings of beer to shrines in their tombs. There are many pictures and sculptures depicting brewing in ancient Egypt (Fig. 1.3). Modern scholars disagree on what can be inferred from these images about the details of ancient Egyptian brewing methods.

Europe

Little is known about the introduction of beer to northern Europe. Historical records from northern Europe before the Middle Ages are incomplete or missing. The **Neolithic** village of Skara Brae in the Orkney Islands off Scotland has yielded what some interpret as evidence of beer brewing 3500–4000 years ago. Finds of possible brewing 3000 miles from Sumer with little in between suggest that Europeans may have invented brewing independently. The Old English epic *Beowulf*, which was written some time around 1000, is set in a heroic Danish culture whose warriors seal their loyalty to their king during elaborate feasting and drinking of beer and **mead**, an alcoholic beverage made from honey.

Monasteries

European **monasteries** played a key role in the development of modern beer. St. Benedict of Nursia (480–547) in Italy wrote a set of monastic rules providing for a daily ration of wine. Beer seems to have been permitted under the rules of St. Gildas (~504–570) in monasteries in Britain and Ireland. St. Columban (~559–615) may have been influenced by Gildas in providing beer for monks in monasteries he founded in France. The monastic customs came together when the synods in 816 and 817 at Aachen brought monasteries in most of Western Europe under a single set of rules. These rules provided that each monk would get a pint of beer or half a pint of wine a day. Monasteries ranged in size from 30 to as many as 400 monks with a similar number of servants and serfs. A monastery that served 150 pints of beer a day would need over 560 gallons (2100 liters) a month. The beer/wine ration assured that many monasteries outside of the grape growing regions would house large breweries. Monasteries served as guest houses for travelers and many sold beer to make extra income. In around 820 a detailed drawing was prepared for renovations of the Monastery of St. Gall in Switzerland. The plan shows three breweries, one near the monks' kitchen, one near the pilgrims' quarters, and one near the guest house. Although there is no indication that the three-brewery plan was actually realized, the St. Gall plan shows that brewing beer had become the norm for a northern European monastery. Starting perhaps in the middle 900s, some monasteries were able to maintain a beer monopoly by controlling the license to produce **gruit**, a mixture of herbs used to flavor beer. This practice waned by the fifteenth century, because **hops** replaced gruit in most regions.

Hops

The hop is a climbing plant whose flowers are used to flavor nearly all beer made today (Fig. 1.4). The first historical record of the use of hops in beer is in a list of rules for monks written in 822. The rules were written by the abbot Adalhard (751–827) for the Monastery of St. Peter and St. Stephen in Corbie, northern France. Adalhard also founded the Corvey Monastery in north central Germany; some sources get these two monasteries mixed up. Hopping of beer at nearby French monasteries in Fontenelle and St. Denis was recorded slightly later. Hops were not cultivated, but were gathered from the wild. The use of hops in beer spread slowly and irregularly throughout Europe. Early evidence of cultivation of hops dates from 859–875 at the Abbey of Freisingen in Bavaria, southern Germany.

The Hanseatic League was a confederacy of trading cities on the north coast of Europe from 1159 to the 1700s. The Hanse traded at North Sea and Baltic

Figure 1.4 Hops on trellis. (See color insert.)

Sea ports from Britain to Russia. One of the major Hanse commodities was beer. In its unhopped form, beer spoils rapidly, making it unsuitable for long distance trading. Hops, in addition to providing a unique flavor to beer, also acts as a preservative. Beer made with hops can stay fresh for weeks or months. The use of hops made beer a transportable commodity, allowing the Hanseatic League to introduce hopped beer to a large region in northern Europe. None of this happened overnight. Powerful people were making good money on gruit, the flavoring used before hops. These people used their influence on taxation and regulation to resist the introduction of a competing flavoring. Added to this is the innate conservatism of people about their food and drink. Hopped beer started to appear in England in the late 1300s mostly for the use of resident foreigners, including officials of the Hanseatic League. Different brewers made unhopped beer, called "ale" and the hopped product, called "beer." By the end of the 1600s all beer in England was hopped. Today we use "beer" as the general term and **ale** is contrasted to **lager** according to the fermentation temperature.

Commerce and Regulation

Starting in the later 1000s, commercial brewers began to set up shop in cities in what is now Belgium. Beer was an ideal product to tax because it was prepared in specialized facilities in batches of fixed size. While it might be possible to make a few pairs of shoes or rolls of wool under the table, it would have been difficult to conceal a batch of beer from the authorities. Beer taxation, both direct and indirect (as by taxing the ingredients), became an important source of revenue for various levels of government. Because of their financial interest in beer, governments got into the habit of regulating the ingredients, preparation, and sale of beer. In addition to taxation, other aspects of the brewing trade were of interest to the town government. Brewing requires heat, which in the Middle Ages meant fire. Breweries were subject to fires that, because of wooden construction, could spread to whole neighborhoods. In an effort to control fire risk, many towns had regulations on where breweries could be built and with what materials of construction. Brewing competes for grain with bread baking, which was seen as essential to feed the population. This may have been the motivation for the famous **Reinheitsgebot** (German: *Reinheit*, purity + *gebot*, order). This regulation, which permitted only barley, hops, and water in beer, was first issued in Munich in 1487. In 1516 the rule was extended to all of Bavaria (southern Germany). One effect of this regulation was a severe limitation on the import of beer into the regions in which it held sway. The *Reinheitsgebot,* in modified form, stayed in effect until it was set aside by the European Union in 1987. Even today it influences brewing practices all over the world.

1.2 THE WORLD OF BEER

Beer Consumption

Today, beer is the most popular alcoholic beverage in much of the world. In 2003, beer consumption in the United States was about 82 liters (231 cans) per adult. This may seem like a lot, but it ranks only eighth among the major beer consuming countries. The beer leader is the Czech Republic at 161 liters (464 cans) per adult. In 2008 the U.S. beer industry had revenues of 23 billion dollars and employed 27,000 workers. That makes beer a small but significant sector of the U.S. economy.

Varieties of Beer

We tend to think of beer in its northern European form. Standard American beer with its light color and clean, uncomplicated flavor belongs to a style called Pilsner lager, which is said to have first been marketed in Pilsen, Bohemia (now in the Czech Republic) on October 5, 1842. There are many styles of beer that are radically different from Pilsner lager. Just to name a few, there is ***chicha*** in

Central and South America, made with maize (corn); **opaque beer** in Africa, made with sorghum; *sake* (sa KEH) in Japan, made with rice; *kvass* in Russia and Eastern Europe, made from bread; and *bouza* in North Africa, made from bread and malt. These products are available in many local variations. Each is made by a unique process, and none tastes anything like Coors Light®.

Beer in Africa

Africa south of the Sahara Desert has an immense diversity of peoples, languages, natural resources, foods, and beer. European style barley beer is brewed and enjoyed in Africa; you may have heard of Tusker Lager®, a product of East Africa Breweries, Ltd. The real story of beer in Africa is in the many local styles using ingredients adapted to growing in the African climate. Beer is made with malt and unmalted flour, often including bananas to provide additional sugar. The grains may be millet (*Pennisetum glaucum*), sorghum (*Sorghum bicolor*), or maize (*Zea mays*, corn). Cloudy beer consumed while it is actively fermenting is called **opaque beer**. Some of these beer styles are produced commercially, for example, SABMiller brews an opaque beer called ***Chibuku***®. *Chibuku* is a maize or sorghum-based beer sold throughout southern Africa, often in waxed paper containers like those sometimes used for milk. A one-liter container costs the equivalent of 40¢. *Chibuku* is yeasty and sour and has a thick layer of sediment that is consumed with the beer. SABMiller characterizes *Chibuku* as "an acquired taste."

In some regions in Africa, beer is more than a beverage and social lubricant. In addition to its use at feasts and parties, beer is the focus of communal work projects. Beer is often perceived as a symbol of prosperity and generosity. Sharing beer can have significance that goes beyond ordinary hospitality. Serving or accepting beer can mark a person's position in the social order. Among the Gamo people of Ethiopia, it is a sign of distinction to be appointed as a *halaka*, a ritual-sacrificer. The appointee must be a wealthy, married, circumcised, morally upright male. Upon appointment, the *halaka* must sponsor two huge feasts at which as many as 300 people come to drink his beer.

Beer in Central and South America

There are many fermented beverages in Central and South America. Beer made from maize (corn) is called *chicha*. The maize can be sprouted and mashed much as Europeans make barley malt beer. Another process used in some areas is to moisten the ground maize with saliva. Enzymes in saliva allow starch from the maize to react with water to give sugar. The treated maize, called *muko*, is dried. When *chicha* is to be made, *muko* and some untreated ground maize are mixed with hot water and converted to sugar. After separation, the sugary liquid is boiled, chilled, and allowed to ferment in clay pots. Various regional styles of *chicha* are made using different varieties of maize and by flavoring the *chicha* with different fruits and spices.

Cauim and *masato* are fermented beverages made from manioc, the starchy root of the cassava shrub (*Manihot esculenta*). The raw roots must be boiled to remove toxic hydrogen cyanide (HCN). The boiled roots are chewed, allowing saliva enzymes to convert the starch to sugar. The chewed roots are boiled, and the sugary liquid is strained, chilled, and fermented.

Beer in the Far East

The best-known (to Westerners) type of beer from the Far East is *sake*, a Japanese beer made from rice (*Oryza sativa*). Related beverages include **huangjiu** from China, and **cheongiu** from Korea.

Sake is made by a unique process. Enzymes to convert the starch to sugar are provided by cultures of mold such as *koji* (*Aspergillus oryzae*), grown on steamed, highly polished rice. Yeast, water, and lactic acid are mixed in to make a starter. More steamed rice is added, and the mixture is allowed to ferment. Additional steamed rice and koji rice are added over a period of several days. At the end of the combined saccharification/fermentation, distilled (pure) alcohol may be added, allowing the product to be diluted with water for a smoother flavor. The *sake* is then pressed out through a filter and pasteurized. Aging can be prolonged.

There are as many styles of rice beer as there are of barley and wheat beer. Rice beer styles differ in the ethanol content, color, conditions of fermentation, degree of polishing of the rice, and sweetness or dryness of the *sake*.

1.3 BEER AND CHEMISTRY

Beer has played an important role in chemistry and biology both from a historical and from a technical point of view. Although many mistakenly attribute the discovery that beer yeast is a living organism to Louis Pasteur (1822–1895), it was reported nearly simultaneously by Charles Cagniard-Latour and Friedrich Traugott Kützing in 1837, and Theodor Schwann in 1839. This observation was strongly attacked by the leading chemists of the time, including Justus von Liebig (1803–1873). The mocking, dismissive tone of their attack seems more characteristic of political talk radio of our own time than scientific discourse. Liebig was eventually driven back from the position that yeast are not living, but to his death he opposed the idea that they are responsible and necessary for alcoholic fermentation. This bitter and fruitful scientific controversy over the nature of fermentation continued between Liebig and Pasteur, ushering in the modern age of biology. Liebig held that fermentation was a nonliving process in which the decaying matter contributed its energy to the breakdown of sugar. Pasteur held that fermentation was a part of the life processes of the microorganisms that were involved with it. The results and arguments of both men were hijacked by followers of the doctrines of **vitalism** and **mechanism**. The vitalists believed that the processes of life could never be explained by the laws that govern ordinary matter. The mechanists believed that living systems follow the same laws as nonliving systems.

Now, 150 years later, we can say that Pasteur was right that all the fermentations that he, Liebig, and anyone else observed were caused by microorganisms. Liebig was right that fermentation is an ordinary chemical process that could, under the right conditions, occur without the participation of living cells. Eduard Buchner (1860–1917) proved this in 1897. Buchner ground up yeast in the presence of abrasives and squeezed out fluid through a cloth. When sugar was added to this fluid, carbon dioxide and alcohol were produced, exactly the same reaction as occurs in live yeast cells. Buchner won the Nobel Prize in chemistry in 1907, the first awarded for a biochemical discovery. In retrospect it is clear that Liebig and Pasteur did not allow themselves to be governed by doctrines like vitalism and mechanism. Pasteur was the first person to apply the principles of microbiology, a field he helped found, to the brewing of beer.

Many advances in chemistry were driven by the needs of the beer industry. These include measurement of temperature and of specific gravity. The Carlsberg Laboratory, set up in Copenhagen in 1875 (one year before Thomas Edison's laboratory in Menlo Park) as an arm of the Carlsberg brewery, was the site of several important discoveries. Emil Christian Hansen (1842–1908) was the first to raise up pure cultures of yeast (or any microorganism) on an industrial scale. The first reliable method of protein analysis was put forward by Johan Kjeldahl (1849–1900). The concept of pH, which is central to water chemistry, was introduced by S. P. L. Sorensen (1868–1939). The Guinness brewery in Dublin gave us the statistical method called **Student's t-test**, invented by William Sealy Gossett (1876–1937).

1.4 ALCOHOL AND PROHIBITION

One of the major reasons that people consume beer is because it contains alcohol. Alcohol, technically termed **ethanol**, is a **psychoactive** substance, which means that it changes the brain function. Depending on the dose, alcohol can lead to anything from mild **euphoria** to **stupor, coma**, and death. For some, alcohol is addictive. The alcohol content of beer, around 3–6%, is low enough so that a thirsty person can drink a glass without, in most cases, a severe effect on his or her ability to function. A similar quantity of wine, which has about three times the alcohol concentration, would be deleterious to one's coordination and judgment. The role of alcohol in society is complex and not altogether positive. One modern issue is the importance of cars in our culture and the problem of driving under the influence of alcohol. Not all such problems are new. Regulations on alcohol are documented in the **Code of Hammurabi** around 3800 years ago in Babylon.

The United States, in its brief history, has embraced and rejected alcohol, often simultaneously. The first English and European settlers in America set up breweries as nearly their first order of business. America offered a new (to Europeans) grain for brewing, **maize** (corn). Major figures in American history, like William Penn, George Washington, and Thomas Jefferson, had breweries attached to their residences.

In the 1800s, opposition to drinking organized into the **temperance movement**. A whole genre of temperance writing arose. The **saloon** or alcohol itself was cast in the role of villain. The progress of the temperance movement came from many sources. These included the aspirations of women for political influence, the rise of a religious movement called **Pietism**, and a fair amount of rowdy drunkenness. Some additional factors began to tip the balance. The arrival of a large number of Irish immigrants engendered a backlash against them and the Catholic religion that many of them practiced. The United States entered the First World War in 1917, which led to hostility toward German-Americans. The temperance movement was able to associate drinking with groups that could be portrayed in a negative light, such as Irish, Catholics, and Germans.

The final factor favoring the temperance advocates was the impending success of the movement to give women the right to vote. Politicians were eager to take positions they thought would appeal to the very large number of women who would soon be voters. Because some of the better-known **prohibition** activists were women, prohibition was seen as a women's issue. Despite the perception at the time, it is not clear that most women were in favor of restricting alcohol. The 18th Amendment to the Constitution establishing prohibition of "intoxicating liquors" was ratified in January 1919. The law that placed prohibition into effect, called the **Volstead Act** (October 1919), surprised many by defining beer as intoxicating liquor. The bill's sponsor, Andrew Volstead (1860–1947, Republican, Minnesota), was defeated in the next election in 1922 after 20 years of service in the House of Representatives.

Prohibition took away a reliable source of tax revenue from all levels of government, and it provided a reliable source of cash for organized crime. Prohibition was expensive to enforce because of widespread resistance to it. The prohibitionists had promised a more sober and productive society with less violent crime. The actual outcome of prohibition seemed at the time to be a loss of respect for tradition and authority characteristic of the Roaring Twenties, and gang violence culminating in the Valentine's Day Massacre in 1929. All of this led to increasing opposition to prohibition. In the 1932 presidential election, both Herbert Hoover and Franklin D. Roosevelt campaigned against prohibition. Roosevelt won by a landslide and was inaugurated on March 4, 1933; on April 7 of that year he signed a law making beer legal. As part of the national celebration, Budweiser introduced its famous Clydesdale horses. The 18th Amendment was repealed on December 5, 1933, and national prohibition came to an end.

Prohibition was not a uniquely American brainstorm. At various times Canada, Russia, Hungary, the Scandinavian countries, and even Australia have flirted with prohibition. Many countries in the Middle East have severe regulations on alcohol designed to restrict its use to visiting foreigners. This is also the situation in Bangladesh and Pakistan. Some states in India are dry. There are still localities in the United States on the scale of counties that prohibit alcohol. Many countries and some states in the United States permit alcohol but have surrounded its production and sale with a nearly impenetrable tangle of regulations.

1.5 BEER TRADITION

Prestige of Beer

Today in Europe and the Americas, beer tends to be seen as a cheap beverage for the lower classes, in contrast to grape wine. Drinking beer is much more likely to be described by unfavorable terms like swilling or guzzling than is drinking wine. Popular images of beer drinking often feature men in undershirts or work clothes; the wine drinkers wear suits and evening gowns. Beer's image is not altogether negative. Beer can be a symbol of peace and friendship. When President Barack Obama met with Professor Henry Louis Gates and police Sergeant James Crowley in the aftermath of a racially charged incident, they drank beer. Nonetheless, beer is much more associated with rowdy sports fans than with presidents. So how did beer go from a beverage considered to be a suitable offering at the tomb of the Pharaoh to its present humble status?

There are two parts to the answer, one historical and one economic. The ancient Mesopotamian heartland of beer was conquered by the wine-drinking Persians 2550 years ago. The Persians were succeeded by the Greeks 2340 years ago, the Parthians 2140 years ago, then the Romans in 116 (1900 years ago). The beer-drinking upper classes of the conquered nations were replaced by wine-drinking conquerors. Class and image are largely determined by what the upper classes do. If the people in power drink wine, then wine is the drink of the powerful and beer belongs to the peasants.

Economically, more expensive, less common items of any sort are regarded as privileges of the upper classes. Wine is more expensive than beer for several reasons. A plot of barley gives much more beer than a plot of grapes gives wine. Grapes must be processed into wine immediately after harvest or they will mold and rot. The winemaker processes the entire harvest of grapes at once. The equipment sits empty for the rest of the year. Grain malt can be stored for months, ready to be brewed into beer at the brewer's convenience. This allows the brewer to keep the brewery operating at high capacity all year, which greatly lowers the cost for a serving of beer. Lower cost is associated with lower prestige.

Role of Beer

Alcoholic beverages, especially beer, are distinct from other foods in having special places to drink with companions. The tavern may be nearly as old as beer itself. Brewing and tavern-keeping were mostly women's trades in ancient times. Taverns often served not only as drinking houses but also as restaurants, hotels, and brothels. If travelers came to town, they would have to go to a tavern. If one wanted to hear stories (true or otherwise) from the outside, one would go to visit the tavern and hoist one with any strangers in attendance. A variety of bars, taverns, saloons, pubs, and beer halls still exist today. It is interesting that some city planners regard good bars as a stabilizing influence in city neighborhoods. They operate late and keep some legitimate traffic in the streets. On the other hand, bad bars bring noise and disorderly behavior with them. A modern complication is dependence on the automobile for travel. Driving a car, especially at night, requires a high level of judgment, perception, and coordination. Driving under the influence of alcohol often has disastrous results and is strictly forbidden all over the world. The problem of how to get home from the bar has resulted in a decrease in alcohol consumption at the tavern in favor of consumption at home. Brewers have responded by providing more beer in cans and bottles, and less in kegs.

Alcoholic beverages are linked to religious and national traditions in ways that go beyond other foods. When people get together, they may raise a glass and recite a ceremonial formula or exchange prayers or good wishes. It would be unconventional to say a toast over a glass of Diet Pepsi®. Western organized religions mostly use wine rather than beer in their rituals, but beer is a key religious symbol among some African and Asian communities. Beer figured very prominently in some ancient religions, including those of Egypt and the Mesopotamian region. An argument could be made that the Super Bowl® is a modern ritual involving beer. Of course, there are also religions that shun and forbid alcohol. These include Islam and some Christian denominations. The use or nonuse of alcohol is thus a characteristic that helps groups define themselves. Beer and other alcoholic beverages have played a key social and religious role since time immemorial, and they can be expected to continue to do so indefinitely.

BIBLIOGRAPHY

Arnold, John P. *Origin and History of Beer and Brewing*. BeerBooks.com, originally published 1911. Some things in this book are no longer accepted today.

Arthur, John W. Brewing Beer: Status, Wealth and Ceramic Use Alteration Among the Gamo of South-Western Ethiopia. *World Archaeology*, **2003**, *34*(3): 516–528. Discusses the social role of beer among an African people.

Bamforth, Charles. *Grape vs. Grain*. Cambridge University Press, 2008. Includes long quotes from translations of the *Hymn to Ninkasi* and the *Kalevala*. Has useful and entertaining information about the science and history of brewing and wine making.

Barnett, James A. Beginnings of Microbiology and Biochemistry: The Contribution of Yeast Research. *Microbiology*, **2003**, *149*: 557–567. The discovery of the living nature of yeast and its role in fermentation. Analyzes the controversy between Pasteur and Liebig.

Behre, Karl-Ernst. The History of Beer Additives in Europe—A Review. *Vegetation History and Archaeobotany*, **1999**, *8*: 35–48. Archaeology of gruit and hops with several maps of the distribution of these herbs and relevant archeological sites.

Buhner, Stephen Harrod. *Sacred and Healing Beers*. Siris, 1998. Discusses ancient and unusual fermented beverages, including *chicha* and *masato*.

Dineley, Merryn; and Dineley, Graham. From Grain to Ale. Skara Brae, a case study. In *Neolithic Orkney in Its European Context*, Anna Ritchie, Ed. McDonald Institute for Archaeological Research, 2001. Evidence of stone age brewing on a Scottish isle 3500–4000 years ago.

Geller, Jeremy. From Prehistory to History: Beer in Egypt. In *The Followers of Horus*, Renée Friedman and Barbara Adams, Eds. Oxbow, 1992. The author participated in the excavation of an ancient Egyptian brewery in Hierakonpolis.

Harper, Philip. *The Insider's Guide to Saké*. Kodansha, 1998. The *sake* story from the only Western *sake* brewer in Japan.

Hein, George E. The Liebig–Pasteur Controversy. *Journal of Chemical Education*, **1961**, *38*: 614–619. Explains the positions and historical background of this controversy.

Horn, Walter; and Born, Ernest. *The Plan of St. Gall*. University of California Press, 1979. Available online at http://www.planstgall.org/horn_born/index.htm. A prime source of information about monasteries in Carolingian Europe. Contains (Vol. III, p. 91) an English translation by Charles W. Jones of Adalhard's *Consuetudines Corbiensis*, source of the first historical mention of hops.

Hornsey, Ian S. *A History of Beer and Brewing*. Royal Society of Chemistry, 2003. Essential reference. Tends to focus on the British isles.

Kramer, Samuel Noah. *History Begins at Sumer*. University of Pennsylvania Press, 1981. Highlights of Sumerian civilization organized by historical "firsts" (first schools, first historian, etc.). The First Pharmacopoea, p. 62, documents the use of beer as a vehicle for administering medicine. This book has no index.

Lagerkvist, Ulf. *The Enigma of Ferment*. World Scientific, 2005. History of scientific understanding of fermentation from ancient times to the award of the 1907 Noble Prize to Eduard Buchner.

McGovern, Patrick E. *Uncorking the Past*. University of California Press, 2009. Highly readable account of the author's work on identification of ancient alcoholic beverages all over the world.

Nelson, Max. *The Barbarian's Beverage*. Routledge, 2005. Top quality scholarship providing critical evaluation of the primary literature of the history of beer in Europe.

Pasteur, Louis. *Studies in Fermentation*, BeerBooks.com, 2005. Originally published in 1876 in French and translated into English by Frank Faulkner in 1879.

van Tangeren, Frank. Standards and International Trade Integration: A Historical Review of the German "Reinheitsbegot." In *The Economics of Beer*, Johan F. M. Swinnen, Ed. Oxford University Press, 2011.

Tremblay, Victor J.; and Tremblay, Carol Horton. *The U.S. Brewing Industry*. MIT Press, 2009. Covers trends and economic factors in recent times. Focus is on large corporate brewers.

Unger, Richard W. *Beer in the Middle Ages and the Renaissance*. University of Pennsylvania Press, 2004. Excellent coverage of the role of monasteries, the spread of hops, commercialization of beer production, international trade in beer, and taxation and regulation of beer in Europe.

Wilson, D. Gay. Plant Remains from the Graveney Boat and the Early History of *Humulus lupulus* in W. Europe. *New Phytologist*, **1975**, *75*(3): 627–648. doi: 10.1111/j.1469-8137.tb01429.x. Correctly translates the first historical mention of the use of hops in brewing. Includes an outstanding summary of the history of hops in Europe.

QUESTIONS

1.1. Where was the earliest chemical evidence of barley beer found? About when was this beer made?

1.2. When was the first historical account of the use of hops in beer?

1.3. What is a monastery? What role did monasteries play in the development of beer?

1.4. What was the Hanseatic League? What role did it play in the development of beer?

1.5. What ingredients were permitted by the *Reinheitsgebot*?

1.6. What was Prohibition? Around when was there national prohibition in the United States?

1.7. What is the psychoactive substance in beer?

1.8. What is opaque beer?

1.9. What is maize?

1.10. What is *chicha*?

1.11. What was gruit?

1.12. What is *sake*?

1.13. Why is beer less expensive than wine?

Fermenter cones. Yuengling Beer Company, Pottsville, Pennsylvania.

CHAPTER 2

WHAT IS BEER?

Beer is a **fermented beverage** made from a source of starch without concentrating the alcohol. The composition of finished beer varies, but the approximate averages for the major components of American commercial beers are about as follows:

BEER COMPOSITION

Water	92.9%
Ethanol	3.9%
Carbohydrates	2.5%
Carbon dioxide	0.5%
Protein	0.2%

This simple analytical result masks a complex reality. Hundreds of minor components not listed above are big players in the flavor and character of the beer. Instead of taking on the issue of what is in finished beer, we will look at how beer becomes beer.

2.1 BEER INGREDIENTS

The usual ingredients for beer are water, malt, hops, yeast, and **adjuncts**. Adjuncts are materials that supplement the malt by providing additional starch or sugar. We will make a quick survey of the ingredients before we cover them in detail in later chapters.

The Chemistry of Beer: The Science in the Suds, First Edition. Roger Barth.
© 2013 John Wiley & Sons, Inc. Published 2013 by John Wiley & Sons, Inc.

Figure 2.1 Barley malt.

Malt

Malt is seeds of grain that are allowed to sprout and are then killed by heating (Fig. 2.1). Most beer malt is made from barley (***Hordeum vulgare***) seeds. Wheat is used for certain styles; oats, rye, sorghum, millet, and others are occasionally used for specialty or regional styles. Malt serves as a source of starch and also provides **enzymes** to break down the starch into sugars that can be fermented.

Barley (Fig. 2.2), among the first plants domesticated as a crop, seems to have originated in what is now the Israel–Jordan area. As a consequence of its importance to civilization, some ancient religions used barley in their rituals. Barley is mentioned many times in religious texts including the Bible.

Each barley plant has one or more stems that, for modern varieties, extend 2 to 3 feet (60 to 90 centimeters). The stem is divided by joints called **nodes**, each of which has a leaf. The flowering head grows from the top node. There are groups of three closely spaced flowers at points along one side of the stem of the flowering head. The next group of three flowers is on the other side of the stem. This gives six rows of flowers. In some varieties of barley, all the flowers are fertile, so after fertilization there are six rows of seeds, called **corn**. These varieties are called **six-row barley**. In other varieties of barley, only the central flower of each group of three is fertile, so there are two rows of corn.

Figure 2.2 Barley. (See color insert.)

These varieties are called **two-row barley**. The barley corn has a groove in the front (ventral) side where it grew against the stem. The corn is covered with interlocking woody shells called **hulls**, one in front and one in the back; the back hull extends to form a characteristic **awn**, also called a beard. Beneath the hulls is a waxy seed coat. Beneath the seed coat is a layer of living cells called the **aleurone layer**. There is a hole in the seed coat, called the **micropyle,** where the hulls meet at the end of the corn away from the awn. The micropyle can admit water. The baby plant, called the **embryo**, is at this end of the corn. The compartment containing the embryo is separated from a compartment containing **starch**, called the **endosperm**, by a divider called the **scutellum**. The endosperm contains granules of starch, each surrounded by a **protein** coat (Fig. 2.3).

Whether the beer is to be made from barley or some other grain, the seeds are first converted to malt, a process called **modification**. The live seeds are soaked in water ("steeped") on and off for about two days. They are then put

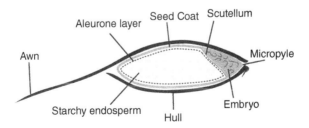

Figure 2.3 Barley corn.

into a box or spread on a floor, provided with moist air, and turned regularly. During this period the embryo wakes up and releases chemical messengers called **hormones**. The hormones direct the cells in the aleurone layer to produce **enzymes**. The enzymes break down the proteins and starch in the endosperm so that the embryo can use them for nourishment until it is able to produce its own food from sunlight. Before this process goes too far, the maltster puts an end to it by heating the germinating seeds in a huge oven called a **kiln**. The temperature is held at 175 to 212 °F (80 to 100 °C). Longer kilning at higher temperatures gives darker, more flavorful grades of malt, but these treatments also destroy a larger fraction of the enzymes.

Hops

The **hop** is a climbing plant, ***Humulus lupulus***. Hop flowers, which look like little pine cones, are used to flavor beer (Fig. 2.4). There are many varieties of hops with names like "Cascade" and "Spalt." All give bitterness, but the different varieties of hops give different flavors. Hops are like wine grapes in the sense that the details of the growing temperature, soil, moisture, and other climate issues affect their flavor. Wine makers deal with variations in grapes by marketing wine as "vintages." Consistency in wine is not expected from one year to the

Figure 2.4 Female hop flowers and leaves. (See color insert.)

next. Brewers do not have this luxury. Some use mixtures of hop varieties so that the proportions can be modified to adjust for local and seasonal variations. Small craft brewers cannot usually keep a big hop inventory on hand, so their product is not as consistent. In the Northern Hemisphere hops are harvested in the late fall and must be dried and cold-stored until used to make beer.

Yeast

Yeast is a single-cell fungus that reproduces by budding. Hundreds of species of yeast have been characterized. Of these, two are usually used for beer making. ***Saccharomyces cerevisiae***, known as **top fermenting** yeast, is used in ale, and ***Saccharomyces pastorianus***, a **bottom fermenting** variety, is used in lager. There are some other species of the genus ***Brettanomyces*** that are used in a Belgian ale style called **lambic**.

Water

You know about water. It turns out that traces of various minerals naturally present in water affect beer in important ways. The chemistry of water is the chemistry of life. Because of its importance, we will devote all of Chapter 4 to water.

2.2 BEER AS FOOD

Ancient beer was cloudy with yeast, protein, and starch and was a nearly complete food. Today's bright clear beer is not as nourishing, but it does contain desirable nutrients like niacin, silicon, and folic acid. Beer has no fat and little or no sugar. Moderate consumption of alcoholic beverages provides some health benefit. Heavy consumption is definitely unhealthful.

Calories in Beer

There is a perception that beer is very fattening. To address this, we need to talk about the primary function of food. Living cells combine food with oxygen to give carbon dioxide, water, and energy for life functions. The overall chemical reaction is the same as burning the food in air, but under controlled conditions and at lower temperature. Extra food that is not needed for the cell's energy requirement is converted to long-term energy storage compounds, including fat. The energy potential in food is determined by burning the food and measuring the amount of heat released. Food energy is usually measured in **Calories** (1 Calorie = 1000 **calories** = 4184 joules = 4 **BTU**). Most people need roughly 2000 Calories a day depending on their size and level of activity.

The Food Energy Potential table (Table 2.1) tells us that the fattening aspect of beer is primarily from the foods that tend to go with it. Two beers, a half dozen wings, and a slice of pizza come to a shocking 1816 Calories. Two beers and a bunch of carrot and celery sticks would come to less than 400 Calories.

TABLE 2.1 Food Energy Potential

Food	Serving	Energy (Cal)
Beer—regular	12 oz (355 mL)	153
Beer—light	12 oz	103
Wine	5 oz	123
Bagel	1	289
French fries	medium	421
Chicken wings	3 wings	603
Pizza	1 slice	304
Carrot	1	25
Celery	1 cup diced	16

Source: U.S. Department of Agriculture: www.nal.usda.gov/fnic/foodcomp/search/index.htm.

Gluten in Beer

Gluten is a mixture of proteins found in the seeds of certain grassy plants, particularly wheat, barley, and rye. The gluten proteins fall into two classes based on solubility: glutelins and prolamines. Wheat has a prolamine called gliadin, barley has hordein, and rye has secalin.

The nutritional issue with gluten is an immune disorder called celiac disease, which affects as much as 1% of the population. Persons with celiac disease react to certain portions of prolamine molecules with a cascade of molecular events leading to damage to the lining of the small intestine. The condition is life threatening and is treated with a diet that excludes all gluten. In addition to persons with celiac disease, many others believe that there is a health benefit to excluding or limiting the intake of gluten. This leads to two questions. Is beer safe for persons with celiac disease? Is beer low enough in gluten to be suitable for persons seeking to limit their gluten intake?

Barley beer has been found to contain traces of hordein fragments of a type associated with celiac symptoms. There is no established safe level of gluten intake for persons with celiac disease. This gives a tentative answer to the first question: no, beer is not known to be safe for persons with celiac disease. Gluten-free beer made without wheat, barley, or rye has been found to be free of traces of celiac-related protein fragments. As for the second question, the brewing process removes more than 99% of the gluten from the barley used to make the beer. Moderate consumption of beer could be regarded as consistent with a low-gluten diet.

Carbohydrates in Beer

Some low-carbohydrate diet books make the incorrect assumption that beer has a high concentration of maltose, a simple carbohydrate that these diets seek to avoid. In fact, the yeast consumes all the maltose and other simple sugars during fermentation. The typical non-light 12 ounce (355 milliliter) beer contains about 12 grams (0.42 ounce) of carbohydrates, a light beer would contain about half of that. In either case, the carbohydrates would be complex, with only traces of simple sugars.

2.3 HOW BEER IS MADE

There are eight basic steps to the standard brewing process, once the raw materials arrive at the brewery. These are milling, mashing, wort separation, boiling, chilling, fermenting, conditioning, and packaging. We will summarize the basics of each step in this brief overview; later chapters will provide a deeper look.

Milling

Malt consists of seeds of grain, usually barley, that have been allowed to sprout to some extent before being killed by heat. During sprouting, proteins called **enzymes** are produced. Later the enzymes will help convert the starch into sugar. The seeds have a packet of starch, plus a small amount of sugar, in a compartment called the **endosperm**. Each seed is surrounded by a **hull** made of a woody material, under which is a **seed coat** made of a waxy material. **Milling** crushes the seeds, breaking the endosperm into small pieces and freeing it from the hull and seed coat so that there is plenty of contact with the water during mashing. In most cases the brewer avoids pulverizing the hull because an intact hull helps in wort separation, and because powdered hulls can give beer a harsh taste or a cloudy appearance. Crushing is done with one to three pairs of rollers in a device called a **mill** (Fig. 2.5). The gap between the rollers defines the size of the particles. Crushed grain is called **grist**.

Mashing

Mashing is the treatment of grist with hot water; it converts most of the starch to sugars that the yeast can use. Mashing is done in a vessel called a **mash tun** (Fig. 2.6). There are three important phases in mashing. The starch particles are **gelatinized**; that is, they absorb water and swell. One reason that barley malt is prized in brewing is because its starch gelatinizes at relatively low temperature compared to other grains. The second phase of mashing is **liquefication**. Starch molecules have long chains of sugar units. **Amylose** has straight chains; **amylopectin** has branched chains. Liquefication breaks these chains into smaller chains called **dextrins**, making the starch more soluble in water. The third phase of mashing, **saccharification**, is the final conversion of starch and dextrins to sugar. Saccharification involves breaking off individual or small groups of sugar molecules from the ends of the starch chains to make sugars that the yeast can use.

In order for liquefication and saccharification to proceed at a reasonable speed, enzymes are needed. Enzymes play the role of **catalysts**, materials that provide a faster pathway for a reaction, but that are not consumed in the reaction. At a higher mash temperature the rate of saccharification is higher, but

Figure 2.5 Malt mill and drive. Parts removed for clarity.

Figure 2.6 Inside the mash tun. Victory Brewing Company, Downingtown, Pennsylvania.

the rate of breakdown of the enzymes, which is called **denaturing**, is also higher. There are two major enzymes important for mashing. One enzyme, **alpha-amylase (α-amylase)**, splits the sugar units at random points in the middle of a chain. The products of alpha-amylase action are shorter starch chains that are not usually usable by yeast, but that provide chain ends for the other enzyme to work on. Alpha-amylase is responsible for liquefication. The other enzyme, **beta-amylase (β-amylase)**, selectively chops off a two-sugar-unit molecule from the end of a starch chain. The two-sugar-unit molecule is a sugar called **maltose**. Beta-amylase is responsible for saccharification. Alpha-amylase is most effective at a higher mash temperature, and beta-amylase is most effective at a lower temperature. The temperature of the mash is adjusted to provide the right balance between the actions of the two enzymes. This was one of the first commercial uses for thermometers.

Also during mashing, some of the proteins from the grist **coagulate**; that is, they solidify like cooked eggs and settle out. Mashing generally takes about one hour.

Wort Separation

After the starch from the grist is converted to sugar and soluble starches, called **dextrins**, the water containing these products (sweet **wort**) must be separated from the spent grains, called **draff**. One method of separation, the lauter process, uses the bed of grain with hulls as a filter. The wort is circulated through the grain bed until it runs clear; then it is drawn off. Hot water is sprayed in to wash all the sugar out of the grain bed, a process called **sparging**. Filtration and sparging can take place in the mash tun itself, or the mash can be transferred to another vessel called a **lauter tun** (Fig. 2.7). Grain hulls are included in the mash to keep the grain bed from forming a tight paste and blocking the flow of wort. The other separation method is mash **filtration**. The mash is driven under pressure through porous polymer pads. The draff can't go through the pores. After wort separation, the spent grain is often sold as cattle feed.

Boiling

The sweet wort is brought to a boil in a vessel called a **kettle** or a **copper** (Fig. 2.8 and Fig. 2.9). Sometimes the heat is delivered by a device called a **calandria**. Hops are usually added at intervals during the boil to provide bitterness and flavor. The wort is boiled for about an hour. Boiling kills unwanted microbes in the wort, coagulates additional protein, and changes molecules from the hop resins to soluble compounds. Boiling also drives off undesirable flavor components. The protein material that coagulates during boiling is called **hot break**. Hot break is usually removed before chilling. Boiling drives off dissolved oxygen and any volatile (easily evaporated) substances that come from the grist or the hops. Sometimes additional hops are added right at the end of the boil, or even after chilling to get aroma compounds from the hops into the beer. The wort is called **hopped wort** after it is boiled.

Figure 2.7 Lauter tun. Victory Brewing Company, Downingtown, Pennsylvania.

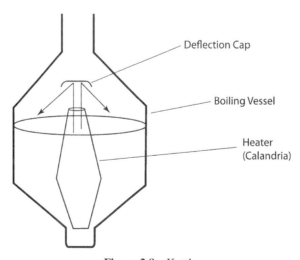

Figure 2.8 Kettle.

Figure 2.9 Kettle. Yuengling Beer Company, Pottsville, Pennsylvania.

Chilling

Heat is removed from the hot hopped wort to cool it down. Cooling generally takes place in a heat exchanger, in which the hot wort flows in tubes or between plates, and a cool liquid flows on the other side of the plates or on the outside of the tubes. Additional protein, called **cold break**, coagulates as the wort cools. Rapid cooling avoids flavor defects. The protein removal during mashing, boiling, and chilling results in clearer beer, which is favored by most beer drinkers. The hops and some or all of the coagulated protein are removed before the wort is fermented. Brewers must be careful to avoid contact of the chilled wort with any source of bacteria or **wild yeasts** (any yeast not chosen by the brewer). Bacterial contamination of the chilled wort can lead to off-flavors, cloudiness, bits of stuff floating about in the beer, and excessive foaming (**gushing**). Fortunately, no known disease-causing organism can grow in beer.

Fermentation

The cool hopped wort is moved into a vessel called a **fermenter** (Fig. 2.10). Oxygen is added to allow the yeast to make compounds that it needs to absorb the wort sugars. Yeast of the desired strain is added ("**pitched**"), and usually the fermenter is sealed off with a one-way valve that allows gas to escape but no air to enter. During the initial stage, lasting a few hours, the yeast adjusts to the fermenter environment by making its membranes more fluid to allow easy intake of sugars from the wort.

In the next stage of fermentation the yeast cells multiply. They take up sugar and produce ethanol, carbon dioxide, and energy for their life processes. The chemical reaction is

Figure 2.10 Fermenter, home brewing. (See color insert.)

$$C_6H_{12}O_6 \text{ (sugar)} \rightarrow 2CO_2 \text{ (carbon dioxide)} + 2\, C_2H_5OH \text{ (ethanol)}$$

a process called anaerobic fermentation. The final product, C_2H_5OH, is ethanol, which makes an alcoholic beverage what it is. In addition to the main reaction, smaller quantities of hundreds of compounds are produced that contribute to the beer's flavor. These contributions vary depending on the strain of yeast and the fermentation temperature. A good deal of heat is released by the fermentation reaction. In a large fermenter, cooling must be provided; otherwise the temperature will rise too high, and the yeast will produce undesired flavors. In a small fermenter, the walls are not so far away, so heat may be able to escape unassisted.

Conditioning

Beer right from the fermenter is called **ruh** (roo) **beer** or **green beer**. The beer must be held for some time in contact with the yeast to achieve a mature flavor

Figure 2.11 Centrifuge. Victory Brewing Company, Downingtown, Pennsylvania.

and to become clearer. For ale, this is a matter of a few days. For lager, **conditioning** can take weeks. Most of the yeast is allowed to settle out.

In commercial brewing, the final step of conditioning is usually filtration or centrifugation to remove any yeast and other particles that did not settle out. Filtration involves passing the beer through a bed of a powdered, highly porous rock called **diatomaceous earth**, also called *kieselguhr*. Diatomaceous earth is formed from the fossilized remains of hard-shelled algae cells called **diatoms** (DIE-uh-toms). Some breweries remove particles by spinning the beer in a device called a **centrifuge** (Fig. 2.11). The centrifuge has a series of rapidly spinning bowls. The spinning causes the heavier particles to settle quickly. The disadvantage of filtration is that the diatomaceous earth yeast mixture has no beneficial use and the brewer has to pay someone to carry it off. The disadvantages of centrifugation are that it can warm up the beer and that the apparatus is very noisy. In some cases, the clarification step is omitted and conditioning continues in the bottle or **cask** after packaging.

Figure 2.12 Kegs. Victory Brewing Company, Downingtown, Pennsylvania.

Packaging

Part of the packaging process is the introduction of carbon dioxide to give the beer its characteristic foam. This can be done by introducing some extra fermentable sugar and allowing live yeast to go into the bottle or cask. A small amount of additional fermentation occurs, producing the desired amount of carbon dioxide. This is called cask or bottle conditioning; it leaves a residue of yeast in the container. The more common way to carbonate beer is to add carbon dioxide under pressure, called **forced carbonation**.

The finished beer can be packaged in convenient-size packages for the consumer, that is, bottles or cans, or it can be put into casks or **kegs** for resale in bars (Fig. 2.12). In all cases, great care is taken to exclude oxygen, which makes the beer stale. Glass bottles should be of brown glass to exclude light, which can introduce a flavor defect called **lightstruck**, a skunky odor due to a reaction with the hop products. Consumer-size packages are usually subjected to some treatment to kill or remove any microbes to make the beer more stable. Heat treatment to kill microbes is called **pasteurization**. Another process is **microbial filtration**. Each of these processes affects the beer's flavor.

Before we can look at the brewing process in detail, we need to work on some of the basics of chemistry.

BIBLIOGRAPHY

Badr, A.; Müller, K.; Schäfer-Pregl, R.; El Rabey, H.; Effgan, S.; Ibrahim, H. H.; Pozzi, C.; Rohde, W.; and Salamini, F. On the Origin and Domestication History of Barley. *Mol. Biol. Evol.*, **2000**, *17*(4): 499–510. Genetic evidence of the origin of domestic barley.

Bamforth, Charles. *Grape vs. Grain*. Cambridge University Press, 2008. Has useful and entertaining information about the science and history of brewing and wine making. Argues that beer should be no less prestigious than wine.

Briggs, D. E. *Barley*. Chapman and Hall, 1978. Everything about barley and how it is grown.

Colgrave, M. L.; Goswami, H.; Howitt, C. A.; and Tanner, G. J. What Is in a Beer? *J. Proteome Res.*, **2012**, *11*: 386–396. Gluten in beer.

Hough, J. S. *The Biotechnology of Malting and Brewing*. Cambridge University Press, 1985. A brief account of the malting and brewing processes.

Palmer, G. H. Cereals in Malting and Brewing. In *Cereal Science and Technology*. G. H. Palmer, Ed. Aberdeen University Press, 1989, p. 61. Microscopic details about barley corn.

Priest, Fergus G.; and Stewart, Graham G. *Handbook of Brewing*. CRC Press, 2006. A detailed coverage of the commercial brewing process. Includes the business side of commercial brewing with 22 pages on pub breweries and microbreweries.

QUESTIONS

2.1. Identify the eight steps in brewing.

2.2. What is grist?

2.3. What are hops? Why are they used in beer?

2.4. About how many Calories are in a serving of beer?

2.5. What quantity does the Calorie measure?

2.6. What issues are associated with gluten in food and drink?

2.7. Explain the function of an enzyme.

2.8. What is the difference between alpha-amylase and beta-amylase?

2.9. What is malt?

2.10. What is a lauter tun? For what is it used?

2.11. Write the chemical equation for the fermentation reaction.

2.12. Give the name and a brief description of each of the three phases of mashing.

2.13. What is pasteurization?

Periodic Table of the Elements

1A																	8A
1 H Hydrogen 1.008	2A											3A	4A	5A	6A	7A	2 He Helium 4.003
3 Li Lithium 6.941	4 Be Beryllium 9.012											5 B Boron 10.811	6 C Carbon 12.011	7 N Nitrogen 14.007	8 O Oxygen 15.999	9 F Fluorine 18.998	10 Ne Neon 20.180
11 Na Sodium 22.990	12 Mg Magnesium 24.305											13 Al Aluminum 26.982	14 Si Silicon 28.086	15 P Phosphorus 30.974	16 S Sulfur 32.066	17 Cl Chlorine 35.453	18 Ar Argon 39.948
19 K Potassium 39.098	20 Ca Calcium 40.078	21 Sc Scandium 44.956	22 Ti Titanium 47.867	23 V Vanadium 50.942	24 Cr Chromium 51.996	25 Mn Manganese 54.938	26 Fe Iron 55.845	27 Co Cobalt 58.933	28 Ni Nickel 58.693	29 Cu Copper 63.546	30 Zn Zinc 65.38	31 Ga Gallium 69.723	32 Ge Germanium 72.61	33 As Arsenic 74.922	34 Se Selenium 78.96	35 Br Bromine 79.904	36 Kr Krypton 83.80
37 Rb Rubidium 85.468	38 Sr Strontium 87.62	39 Y Yttrium 88.906	40 Zr Zirconium 91.224	41 Nb Niobium 92.906	42 Mo Molybdenum 95.96	43 Tc Technetium [98]	44 Ru Ruthenium 101.07	45 Rh Rhodium 102.906	46 Pd Palladium 106.42	47 Ag Silver 107.868	48 Cd Cadmium 112.411	49 In Indium 114.818	50 Sn Tin 118.710	51 Sb Antimony 121.760	52 Te Tellurium 127.60	53 I Iodine 126.904	54 Xe Xenon 131.29
55 Cs Cesium 132.906	56 Ba Barium 137.327	57 La Lanthanum 138.906	72 Hf Hafnium 178.49	73 Ta Tantalum 180.948	74 W Tungsten 183.84	75 Re Rhenium 186.207	76 Os Osmium 190.2	77 Ir Iridium 192.22	78 Pt Platinum 195.084	79 Au Gold 196.967	80 Hg Mercury 200.59	81 Tl Thallium 204.383	82 Pb Lead 207.2	83 Bi Bismuth 208.980	84 Po Polonium [209]	85 At Astatine [210]	86 Rn Radon [222]
87 Fr Francium [223]	88 Ra Radium [226]	89 Ac Actinium [227]	104 Rf Rutherfordium [265]	105 Db Dubnium [268]	106 Sg Seaborgium [271]	107 Bh Bohrium [270]	108 Hs Hassium [277]	109 Mt Meitnerium [276]	110 Ds Darmstadtium [281]	111 Rg Roentgenium [280]	112 Cn Copernicium [285]						

58 Ce Cerium 140.166	59 Pr Praseodymium 140.908	60 Nd Neodymium 144.242	61 Pm Promethium 144.91	62 Sm Samarium 150.36	63 Eu Europium 151.965	64 Gd Gadolinium 157.25	65 Tb Terbium 158.925	66 Dy Dysprosium 162.50	67 Ho Holmium 164.930	68 Er Erbium 167.26	69 Tm Thulium 168.934	70 Yb Ytterbium 173.054	71 Lu Lutetium 174.967
90 Th Thorium 232.038	91 Pa Protactinium 231.036	92 U Uranium 238.029	93 Np Neptunium [237]	94 Pu Plutonium [244]	95 Am Americium [243]	96 Cm Curium [247]	97 Bk Berkelium [247]	98 Cf Californium [251]	99 Es Einsteinium [252]	100 Fm Fermium [257]	101 Md Mendelevium [258]	102 No Nobelium [259]	103 Lr Lawrencium [262]

CHAPTER 3

CHEMISTRY BASICS

3.1 ATOMS

The central idea of chemistry is that everything in the world, including beer, is made of tiny particles called **atoms**. A typical atom has a diameter of about 0.2 nanometer (8 billionths of an inch). The smallest object visible to the naked eye is about 0.1 millimeter (0.004 inch). A speck of carbon this size would contain about 60,000,000,000,000,000 atoms.

There are around a hundred known types of atoms; each type represents a particular **element**. Atoms consist of a central particle called the **nucleus** surrounded by a cloud of very light particles called **electrons** (Fig. 3.1). The nucleus has a positive electrical charge; the electrons have a negative charge. The positive charge on the nucleus is due to positively charged particles called **protons**. The number of protons that an atom has in its nucleus is called its **atomic number**. The atomic number determines what element it is. The most important elements in brewing beer are hydrogen (1 proton), carbon (6 protons), nitrogen (7 protons), and oxygen (8 protons). The nuclei of most elements also contain uncharged particles called **neutrons**. Normally an atom has the same number of electrons as protons. The charge on the electron and proton are equal but opposite in sign. As a result, the atom is electrically **neutral**; that is, it has no charge. The protons and neutrons in an atom are fixed; none are gained or lost in a chemical reaction. The electrons are much smaller, lighter, and less tightly **bound** than the protons, so it is not unusual for electrons to be transferred from one atom to another, or even to be shared by two atoms.

The Chemistry of Beer: The Science in the Suds, First Edition. Roger Barth.
© 2013 John Wiley & Sons, Inc. Published 2013 by John Wiley & Sons, Inc.

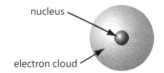

Figure 3.1 Atom structure.

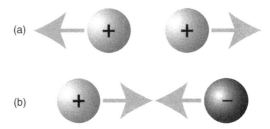

Figure 3.2 (a) Same signs—repel; (b) opposite signs—attract.

The forces that pull atoms together or push them apart are the forces between electrical charges. Particles with charges that have the same sign (both positive or both negative) push one another away; they **repel** one another. Particles with charges of the opposite sign (one positive and the other negative) pull together; they **attract** one another (Fig. 3.2). The reason that electrons stay at the atom is because electrons have a negative charge and the nucleus has a positive charge. Neutrons, being neutral, exert no forces on electrons or on other atoms, so they are not usually important to the chemistry of beer.

The atoms are often represented by symbols, such as H for hydrogen, C for carbon, N for nitrogen, O for oxygen, and Na for sodium. Symbols for elements are one or two letters; the first letter is capitalized and the second (if any) is lowercase. Not all symbols correspond to the English name of the element. The symbols for elements are the same in all languages. Even languages that don't use our Latin characters use the same symbols. Some elements that are involved in beer and brewing are listed with their symbols, atomic numbers, and molar masses (more later) in Table 3.1.

3.2 ENERGY LEVELS AND THE PERIODIC TABLE

The electrons in atoms are arranged in layers called **shells**. The energy of the electrons increases as they go into shells that are farther from the nucleus. For this reason, the shells are also called **energy levels** (Fig. 3.3). The first (closest to the nucleus) shell can have no more than two electrons. The

TABLE 3.1 Elements and Symbols

Element	Symbol	Atomic Number	Molar Mass (g/mol)
Hydrogen	H	1	1.008
Carbon	C	6	12.011
Nitrogen	N	7	14.007
Oxygen	O	8	15.999
Sodium	Na	11	22.990
Magnesium	Mg	12	24.305
Aluminum	Al	13	26.982
Phosphorus	P	15	30.974
Sulfur	S	16	32.066
Chlorine	Cl	17	35.453
Calcium	Ca	20	40.078
Iron	Fe	26	55.845
Copper	Cu	29	63.546

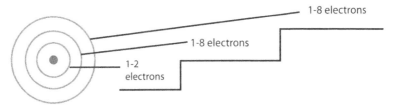

Figure 3.3 Energy levels.

second two shells can have no more than eight electrons each. The closer, lower energy shells fill completely before any electrons go into farther, higher energy shells. Sodium has a total of 11 electrons, so two electrons fill the first shell, eight fill the second shell, and the one remaining electron occupies the third shell. Carbon has six electrons, two in the first shell and four in the second. The highest or outermost occupied shell is the one that interacts with nearby atoms and defines the behavior of an atom. This shell is called the **valence** shell. In sodium, the valence (third) shell has one electron; in carbon the valence (second) shell has four electrons. The nucleus plus electrons in shells of lower energy than the valence shell are collectively known as the **core**. Core electrons are not usually gained, lost, or shared. It is convenient to think of an atom as a positively charged core consisting of the nucleus plus the core electrons, surrounded by the chemically significant valence electrons.

The **periodic table** of the elements arranges the elements in rows, called **periods** from left to right in order of atomic number organized in columns called **groups** that reflect the number of valence electrons. There is a periodic table facing the opening page of this chapter. Sodium is in group 1A, carbon is in group 4A, nitrogen is in group 5A, and oxygen is in group 6A. The groups identified as 1A to 8A are the **main groups**. In the main groups, the group number is the same as the number of valence electrons.

Elements in a group all have the same number of valence electrons, so there are certain similarities in their behavior. For example, elements in group 8A are very unreactive; they are called **noble gases**. The elements in the lower and left portion of the periodic table are classified as **metals**. Their valence electrons tend to be loosely held. Metals have **metallic luster** (chemistry jargon for shiny), and they are good conductors of electricity and heat. The elements in the upper and right portion of the periodic table plus hydrogen are classified as **nonmetals**. They have tightly held electrons and are usually poor conductors of electricity and heat. In between are a few elements that behave like metals in some ways and like nonmetals in others. These are called **semimetals** or **metalloids**.

3.3 COMPOUNDS

Atoms can combine in various ways to make **compounds**. A compound is a substance in which atoms of more than one element are held together by forces called chemical **bonds**. There is a distinction between a compound, in which the atoms are held together by chemical bonds, and a **mixture**, in which the components (which could be elements or compounds) are not bonded together. To illustrate this distinction we could take some hydrogen, an element whose molecules have two atoms, each of which has one proton, and mix it with oxygen, which has diatomic (2-atom) molecules of 8-proton atoms, and we would get a mixture. There would be two different substances in our container. The properties of the mixture would derive from the individual properties of the components of the mixture, hydrogen and oxygen. Because hydrogen and oxygen are both colorless gases, a mixture of hydrogen and oxygen is also a colorless gas. The mixture could sit indefinitely if nothing happens to cause a chemical reaction. Now suppose we introduce a spark into the container. The hydrogen and oxygen undergo an explosive chemical reaction that releases energy. When things settle down, we find that the hydrogen and oxygen are gone. All that remains is some steam and a few drops of water sticking to what is left of the walls of the container. Water, a compound of hydrogen and oxygen, is not at all like hydrogen or oxygen. At room temperature it is a liquid. It is made up of molecules whose structure is H–O–H. The hydrogen atoms are no longer attached to one another; they are now attached to the oxygen atoms. The oxygen atoms are attached to the hydrogen atoms. Every single atom that we started with is still there, but they are now combined in different ways to give water, a compound containing hydrogen and oxygen atoms. Prior to the spark-initiated reaction, this compound was not present. It was produced by the rearrangement of the hydrogen and oxygen atoms in the reaction.

COMPOUNDS   37

One general principle in compound formation is that an atom of a main group element tends to react in ways that give it either a filled or an empty valence electron shell. This explains why atoms from group 8A on the periodic table, the noble gases, are unreactive; their valence shells are already filled. There are two main ways that an atom can fill or empty its valence shell. The atom can gain or lose electrons, forming a charged particle called an **ion**, or it can share electrons with other atoms. The shared pair of electrons holds the atoms together by forming an attachment called a **covalent bond**.

Ions

Sodium can empty its valence shell by giving one electron to another atom and taking a charge of +1. Oxygen can fill its valance shell by taking two electrons from other atoms and taking a −2 charge (Fig. 3.4). An atom that has gained or lost an electron so it has an electrical charge is called an **ion**. The

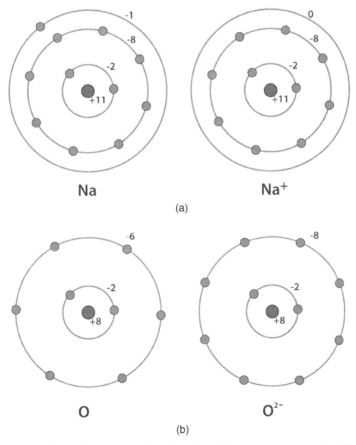

Figure 3.4 (a) Sodium atom and sodium ion; (b) oxygen atom and oxide ion.

38 CHEMISTRY BASICS

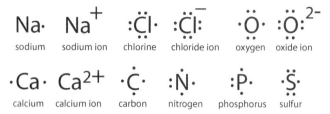

Figure 3.5 Lewis dot diagrams.

charge on an ion is always shown as a superscript: the formula for the sodium ion is Na^+, the formula for the oxide ion is O^{2-}. The charge is part of the formula of an ion; it is not optional.

There is a very big difference between an element and its ion. The element chlorine exists as a two-atom molecule, Cl_2. It is a pale green, highly toxic gas that is used at low concentration to disinfect water. The chloride ion, which is a chlorine atom that has gained an electron, is found in ionic compounds like sodium chloride (NaCl), table salt.

The charges on ions of elements in groups 1A, 2A, and 3A are usually equal to the group number, so sodium (group 1A) forms the Na^+ ion and calcium (group 2A) forms the Ca^{2+} ion. Atoms in groups 6A and 7A often form negative ions with a charge of the group number minus 8. Oxygen is in group 6A, $6 - 8 = -2$, so the oxide ion is O^{2-}. Chlorine is in group 7A, $7 - 8 = -1$, so the chloride ion is Cl^-.

Lewis Dot Diagrams

A symbol of an element or ion with a dot for each valence electron is called a Lewis dot diagram (Fig. 3.5), after chemist G. N. Lewis (1875–1946). The Lewis dot diagram helps us understand ion formation and electron sharing in the main group elements. These diagrams form the basis of Lewis structures, which are the standard way to show the structures of molecules.

3.4 IONIC BONDS

An **ionic bond** is the force of attraction between ions (charged particles) of opposite charge. Sodium is in group 1A; it can give up its one valence electron to empty its valence shell and become the positively charged sodium ion (Na^+). Chlorine is in group 7A; it can gain one electron to fill its valence shell with eight electrons and become the negatively charged chloride ion (Cl^-). These ions clump together in a way that gets the positive and negative ions as close together as possible while keeping the ions of like charge apart. The result is a three-dimensional arrangement called a **crystal**. In Figure 3.6, the larger balls

Figure 3.6 Ionic bonding, sodium chloride crystal. Photo by Roger Barth. (See color insert.)

represent chloride ions and the smaller balls represent sodium ions. Compounds that result from ionic bonding do not have individual molecules. All the ions in a crystal are bound together.

The rule for ionic compounds is that there must be the same amount of positive charge as negative charge to give the compound a net charge of zero. In the case of sodium chloride, there must be one chloride ion for every sodium ion. The compound made from calcium ions (Ca^{2+}) and chloride ions (Cl^-) needs two chloride ions for each calcium ion. We express the elemental makeup of this compound in symbols as $CaCl_2$, which is an example of a chemical **formula**. A formula is a representation of a compound using the symbols for the elements in it. The number of atoms of each element is shown as a subscript to the right of the symbol for that element (a subscript of one is omitted). The charges of the ions are not shown in the formula for an ionic compound. An easy way to find the formula of an ionic compound is the criss-cross rule. To use the criss-cross rule, write the positive and negative ions. Then make the charge of each ion (omitting the sign) the subscript of the other ion, as shown in Figure 3.7.

Some ions have several atoms bound together by covalent bonds (electron sharing). These ions are called **polyatomic ions** (see Table 3.2). Even though there are covalent bonds within the polyatomic ion, the compounds they form are ionic. The formula of the compound made from sodium ions (Na^+) and the phosphate ion (PO_4^{3-}) is Na_3PO_4, called sodium phosphate.

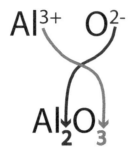

Figure 3.7 Criss-cross rule.

TABLE 3.2 Polyatomic Ions

Formula	Name
NH_4^+	Ammonium ion
CO_3^{2-}	Carbonate ion
HCO_3^-	Bicarbonate ion
OH^-	Hydroxide ion
NO_3^-	Nitrate ion
NO_2^-	Nitrite ion
O_2^{2-}	Peroxide ion
O_2^-	Superoxide ion
PO_4^{3-}	Phosphate ion
HPO_4^{2-}	Hydrogen phosphate ion
$H_2PO_4^-$	Dihydrogen phosphate ion
SO_4^{2-}	Sulfate ion
HSO_4^-	Bisulfate ion
SO_3^{2-}	Sulfite ion

3.5 COVALENT BONDS AND MOLECULES

Apart from gaining or losing electrons, an atom can get a filled valence shell by sharing electrons with another atom, so both atoms get to use the shared electrons. A hydrogen atom, which has one electron, needs one more electron to fill its valence shell. If two hydrogen atoms each share their one electron, each will have a filled valence shell of two electrons, as shown in Figure 3.8.

A shared pair of electrons is called a **covalent bond**. Covalent bonding is common among atoms of nonmetal elements in the upper right of the periodic table, plus hydrogen. A **molecule** is a group of atoms held together by one or more covalent bonds. The bonds between the atoms in a molecule are much stronger than any forces between atoms of different molecules. In the case of hydrogen, the two hydrogen atoms are held together by a covalent bond, so they travel together as a two-atom molecule. The electron pair is often shown as a line between the symbols of the atoms: H–H. A diagram showing a molecule in this way is called a **structural formula**. The number of covalent bonds that an atom makes to fill its valence shell is called its **valency**. The valency of

Figure 3.8 Covalent bond.

TABLE 3.3 Charges and Valencies

Element	Symbol	Valence Electrons	Usual Charge	Usual Valency
Hydrogen	H	1	+1	1
Carbon	C	4	None	4
Nitrogen	N	5	−3	3
Oxygen	O	6	−2	2
Sodium	Na	1	+1	1
Magnesium	Mg	2	+2	2
Aluminum	Al	3	+3	3
Phosphorus	P	5	−3	3, 5
Sulfur	S	6	−2	2, 6
Chlorine	Cl	7	−1	1
Calcium	Ca	2	+2	2
Iron	Fe	8	+2, +3	6
Copper	Cu	11	+1, +2	6

hydrogen is 1; the usual valency of most other main-group nonmetals is [8 − group number]. Some atoms in the third through last rows of the periodic table can also take a valency equal to the group number. For example, phosphorus, group 5, usually has a valency of 3, but in some compounds it takes a valency of 5. Atoms in the first two rows of the periodic table cannot take a valency higher than 4. (See Table 3.3.)

The valencies of the atoms control the structures of the molecules they form. An atom with a valency of 1 makes one bond. That means it must be on the end or outside part of the molecule because it can only attach in one place. Higher valency atoms can be in the middle of the molecule. Oxygen has a valency of 2; hydrogen has a valency of 1. The simplest compound of oxygen and hydrogen has two H atoms bound to one O atom, giving H–O–H, water. It is also possible to have two oxygen atoms bound together with one hydrogen atom at each end, giving H–O–O–H, hydrogen peroxide. Is there a compound with a chain of three oxygen atoms, H–O–O–O–H? This molecule, trioxidane, has been reported, but it is so unstable that special low-temperature techniques are needed to isolate tiny quantities of it. It turns out that for most elements, chains of atoms become less stable as the chains grow longer. The key exception is carbon, which can form stable chains without limit.

Some atoms can share more than one pair of electrons. This is only common in atoms of C, N, O, S, and P. Two shared pairs make a **double bond**. Three shared pairs make a **triple bond**. Double bonds are shown as two lines; triple bonds as three. A compound of carbon (valency 4) and oxygen (valency 2) is O=C=O, carbon dioxide.

NAMES OF CHEMICAL COMPOUNDS

Ionic Compounds

Compounds with a metal and a nonmetal or with any polyatomic ion are named as ionic compounds. The name consists of the name of the positive ion, followed by that of the negative ion. Monoatomic (one-atom) ions are named by the element from which they derive. In the case of negative ions, the suffix -ide is added to the stem of the name. The stem is the name without suffixes like -ine, -ogen, -ygen, or -ur. The ion corresponding to oxygen is the oxide ion: K_2O is potassium oxide. Many metals that are not from the main groups (and even a few that are) can take more than one charge. In these cases the charge, in the form of a Roman numeral, is added to the name of the element: $Fe(NO_3)_2$—iron(II) nitrate. Polyatomic ions (ions with more than one atom) have an irregular pattern of names.

Binary (Two-Element) Compounds of Nonmetals

The number of each atom in the formula is indicated by prefixes representing Greek numbers. Often mono- (one) is omitted. Before a vowel, the a- is omitted. The second element named gets an -ide suffix to the stem: for example, N_2O_5—dinitrogen pentoxide. The number prefixes should not be used for ionic compounds, although sometimes they are in common names. Many binary compounds of nonmetals have well-established common names. H_2O is called water, not dihydrogen oxide. NH_3 is called ammonia.

NUMBER PREFIXES

1	mono-
2	di-
3	tri-
4	tetra-
5	penta-
6	hexa-
7	hepta-
8	octa-
9	nona-
10	deca-
11	undeca-
12	dodeca-

Acids

Compounds of hydrogen and a group 7A element have the annoying characteristic of having two names. In the absence of water, they are named as binary compounds of nonmetals. Anhydrous (no water) HCl is called hydrogen chloride. When dissolved in water, they are named as hydroacids. They take the prefix hydro- followed by the stem of the group 7A element with an -ic acid suffix. Aqueous (in water) HCl is called hydrochloric acid.

Compounds that can be thought of as being the result of adding enough H^+ ions to balance the charge on a polyatomic ion containing oxygen are called oxyacids. Oxyacids are named by changing the -ate suffix on the polyatomic ion to -ic acid. If the polyatomic ion has an -ite suffix it changes to -ous acid. NO_3^- is nitrate ion; HNO_3 is nitric acid. SO_3^{2-} is sulfite ion; H_2SO_3 is sulfurous acid (notice that we put the -ur suffix back on sulfur).

3.6 MOLECULAR SHAPE

Every molecule has a specific shape that gives it the lowest possible energy. The bonds that define the positions of the atoms point in specific directions and have specific lengths, so we can figure out approximately what shape a molecule will have. The **underlying geometry** of an atom in a molecule is governed by the groups of valence electrons on that atom. These groups repel one another because, like all electrons, they have negative charges. An **electron group** can be an unpaired electron (rare), a pair of unshared electrons, or a covalent bond. One covalent bond—single, double, or triple—counts as one electron group. If there are two electron groups on an atom, they occupy positions directly across the atom from one another, so the group, the central atom, and the second group form a straight line. This arrangement is called a **linear** underlying geometry (Fig. 3.9). An example would be O=C=O, carbon dioxide. Carbon has two electron groups, the two double bonds. The carbon dioxide molecule is linear; that is, the three atoms fall on a straight line.

If an atom has three electron groups, they take positions forming a triangle evenly spaced about the central atom. The angle formed by a group, the central atom, and another group is (ideally) 120 degrees (Fig. 3.10). This geometry is

Figure 3.9 Linear geometry.

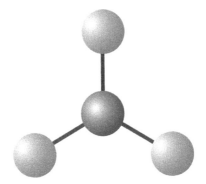

Figure 3.10 Trigonal planar geometry.

Figure 3.11 Formaldehyde.

called **trigonal** (TRIG un ul) **planar** because the three groups and the central atom form a flat plane. In formaldehyde the carbon has a double bond to the oxygen atom (one group) and two bonds to hydrogen atoms (one group each) for a total of three electron groups (Fig. 3.11). The ideal geometry is trigonal planar with 120 degree angles; the actual structure of formaldehyde has a slightly smaller H–C–H angle (116.5 degrees) and slightly larger H–C=O angles (121.7 degrees). Deviations from the ideal angles are common when the atoms associated with the electron groups are not the same.

Four electron groups take positions in three dimensions with angles of about 109.5 degrees (Fig. 3.12). This geometry is called **tetrahedral** (tet ra HEED rul), because if we were to draw lines connecting the four electron groups, the resulting solid would be a tetrahedron. Methane, CH_4, has four single bonds on the central carbon atom. The molecule is tetrahedral. The tetrahedral underlying geometry is the most common geometry in molecules important for beer.

In the examples above, the electron groups are covalent bonds. Electron groups can also be unshared pairs. In these cases there is a distinction between the underlying geometry and the **shape**. The shape considers only atoms and bonds, not the unshared pairs. Sulfur dioxide, sometimes used as a preservative in beer and wine, has a central sulfur atom with a double bond (one group) to oxygen, a single bond (one group) to the other oxygen, and an unshared pair (one group) for a total of three electron groups. An atom with three electron

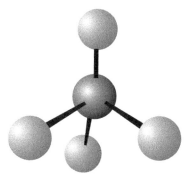

Figure 3.12 Tetrahedral geometry.

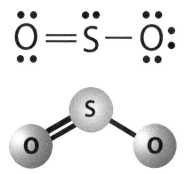

Figure 3.13 Bent shape.

groups has a trigonal planar underlying geometry, whose ideal angles would be 120 degrees. Because one of the electron groups is an unshared pair, the shape consists of a sulfur atom with two oxygen atoms with an O–S=O angle of 119.6 degrees. The molecule has a bent shape (Fig. 3.13).

Nitrogen is in group 5A and has a valency of 3; its simplest compound is NH_3, ammonia. The nitrogen in ammonia has five valence electrons, three of which are shared with electrons from hydrogen to form three bonds. The remaining two electrons are unshared. So the nitrogen has a total of four electron groups, three bonds plus an unshared pair. Four electron groups give a tetrahedral underlying geometry. The three bonds form a tripod shape with angles of 107 degrees (Fig. 3.14). The hydrogen atoms are on one side and the nitrogen atom is on the other.

A molecule with four electron groups, two of which are bonds and the other two of which are unshared pairs, would have a bent shape with an angle of (ideally) 109.5 degrees. Water has a central oxygen atom with two bonds to hydrogen and two unshared pairs for a total of four electron groups. Two of

46 CHEMISTRY BASICS

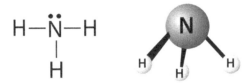

Figure 3.14 Tripod shape.

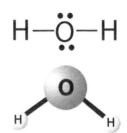

Figure 3.15 Bent shape.

Figure 3.16 Polar bond.

these are bonds, so the molecule is bent (Fig. 3.15). Water has an oxygen side and a hydrogen side. The actual bond angle in water is 104.5 degrees.

3.7 POLARITY AND ELECTRONEGATIVITY

The electrons in a covalent bond are shared, but they are not always shared equally. Atoms of different elements differ in their tendencies to pull the shared electrons in a covalent bond toward themselves. The tendency for an atom to attract the bonding electrons is called **electronegativity**. Atoms of metals and elements in group 8A (the noble gases) have low electronegativities. Electronegativity increases as one goes upward and to the right (but no further than group 7A) on the periodic table. When atoms of different electronegativities share electrons to form a covalent bond, the electrons will be more attracted to the more electronegative atom, giving it a partial negative charge and leaving the other atom with a partial positive charge (Fig. 3.16). A covalent bond with a negative and a positive end is said to be **polar**. The greater the difference in electronegativity, the more polar the bond. In the most extreme cases, the electron is completely transferred to the more electronegative atom and the bonding is ionic. Unequal sharing gives a continuous range

POLARITY AND ELECTRONEGATIVITY

TABLE 3.4 Electronegativities

H	2.2
C	2.5
N	3.0
O	3.5
F	4.0
P	2.2
S	2.6

Figure 3.17 Polar molecule.

Figure 3.18 Polarity cancellation.

of bonding character from essentially covalent, when the atoms are of equal electronegativity, to predominantly ionic, when the electronegativity difference is greater than 1.7. In between are polar bonds with some ionic and some covalent character. It is worth remembering that C–H bonds are nearly nonpolar; C–N and N–O bonds are moderately polar; and O–H, C–O, C=O, and N–H bonds are highly polar. (See Table 3.4.)

If the whole molecule has a negative and a positive end, as in Figure 3.17, the molecule is polar. The degree of polarity of a molecule determines how the molecule will interact with other molecules. Most polar molecules have at least one polar bond. A molecule can have polar bonds and still not be polar. Carbon dioxide has two C=O double bonds. Each of these is polar; the electrons are attracted to the oxygen end, making it more negatively charged than the carbon end. Because the carbon atom has two electron groups (two double bonds) the geometry and shape are linear; the oxygen atoms are 180 degrees apart on opposite sides of the carbon atom. One oxygen atom pulls electrons to the right and the other pulls them to the left, so the polarities of the bonds cancel one another out (Fig 3.18). The cancellation makes carbon dioxide nonpolar. Water, by contrast, has a bend that brings the polar hydrogen–oxygen bonds to one side of the molecule, so water is a polar molecule.

3.8 INTERMOLECULAR FORCES

Whether a substance is a solid, liquid, or gas; whether it will dissolve in water, oil, or not at all; whether it is strong or weak; tough or brittle, all depend on the forces that attract the individual particles to one another. For molecular substances, these forces are called **intermolecular forces**.

Dispersion Forces

A molecule can be thought of as a positively charged framework of atomic cores surrounded by a negatively charged cloud of valence electrons. The positive framework is stiff and difficult to deform. The negative valence electron cloud is loosely held and can move more easily. If the molecule is polar, the center of the positively charged framework and that of the negatively charged valence electron cloud are at different points. If the molecule is not polar, the *average* positions of the positive and negative centers are at the same point, but the negative center can move about as the electrons shift in their orbits. So at any given time, there can be a slight temporary polarity even in a nonpolar molecule (Fig. 3.19). When a molecule is temporarily polar, nearby molecules can respond because their electron clouds will shift to get closer to the positive end and farther from the negative end of their temporarily polar neighbors. This gives rise to a net force of attraction called the **dispersion** force. Dispersion forces tend to be stronger in larger molecules and in molecules with more loosely held electrons, which are more easily shifted out of their average positions. Molecules that are long and thin have more opportunities to interact, so their dispersion forces are stronger than those in short thick molecules. All molecules have dispersion forces. In the absence of hydrogen bonding (see below) dispersion forces are usually the dominant force between molecules.

Stacking Forces

Certain molecules that take the form of a ring have loosely held clouds of electrons above and below the ring. These rings can stack like dishes and expose a large amount of surface area to interactions with one another. The forces between stacked rings can be stronger than ordinary dispersion forces. **Stacking forces** are important in the strength of Kevlar®, a plastic that is used in bulletproof vests, and in many biological compounds including proteins and

Figure 3.19 Nonpolar, temporary polarity.

INTERMOLECULAR FORCES   49

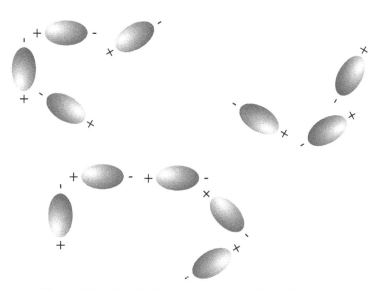

Figure 3.20 Gallic acid (tannin).

Figure 3.21 Dipole–dipole interactions. (See color insert.)

DNA. Stacking forces give rise to some of the properties of tannins that come from grain hulls and hops and that can make beer hazy (Fig. 3.20).

Dipole–Dipole Interactions

Polar molecules tend to align themselves so the ends with opposite charges come close and the ends with like charges stay apart (Fig. 3.21). As a result, there is a force of attraction between the molecules called the **dipole–dipole interaction**. The size of the force depends on the magnitude of the polarity of the molecules. Because of the dipole–dipole interaction, polar molecules stick to one another more strongly than they stick to nonpolar molecules.

50   CHEMISTRY BASICS

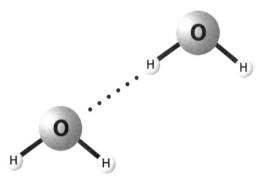

Figure 3.22   Hydrogen bonding.

Hydrogen Bonding

Hydrogen atoms bonded to the highly electronegative atoms nitrogen (N), oxygen (O), and fluorine (F) can have a unique interaction called **hydrogen bonding** (Fig. 3.22). The more electronegative atom attracts the shared electrons to itself. Because a hydrogen atom only has one electron, this atom is now stripped of electrons and is a virtually bare proton. The bare proton has a very high density of positive charge that feels a strong force of attraction to any unshared pairs of electrons, like those on another N, O, or F atom. This attraction is hydrogen bonding. What makes hydrogen bonding distinct from the ordinary attractions of dipoles is that, in addition to being somewhat stronger, a hydrogen bond has directional character. The H–O–H bond angle for hydrogen bonding is about the same as the ideal tetrahedral angle of 109.5 degrees.

In ice, water molecules form 12-member rings of alternating O and H atoms. Every other bond is a hydrogen bond. Because of the large cavity in the ring, ice is less dense than liquid water, hence it floats. Hydrogen bonding is important in determining the structure of water, sugars, starches, alcohols, proteins, and DNA.

Hydrophobic Force

Because of the strong forces attracting polar molecules, or polar parts of molecules to one another, they tend to squeeze nonpolar molecules or parts of molecules out of regions that they occupy. The nonpolar molecules are driven together and seem to avoid polar molecules. In watery materials, like beer or the fluid inside cells (cytoplasm), the hydrogen-bonding water molecules drive nonpolar molecules together. The water and the nonpolar molecule or region seem to repel one another. This apparent repulsion is called the **hydrophobic force**. We don't need to introduce a new force to account for this, but the concept of a hydrophobic force can be useful in systems dominated by water. In this context, polar and hydrogen bonding molecules or regions are said to be **hydrophilic** [Greek: *hydor*, water + *philein*, love] and nonpolar molecules or regions are said to be **hydrophobic** [Greek: *hydor*, water + *phobos*, dread].

Polar molecules have a strong force of attraction to one another. Any nonpolar molecules among them get pushed aside to allow the polar molecules to interact. As a result, polar and hydrogen-bonding molecules or parts of molecules tend to stay away from nonpolar molecules or parts of molecules. That is why ethanol, a polar, hydrogen-bonding molecule, can dissolve easily in water, but oxygen, a nonpolar molecule, has very limited solubility. This effect is described in the phrase "like dissolves like." Cell membranes, like those in yeast cells, have a nonpolar layer that resists the passage of ions and polar molecules. When a protein needs to make a selective binding site for a particular molecule, it can offer a nonpolar pocket to attract the nonpolar parts, polar groups like C=O to attract polar parts, and OH or NH groups to attract hydrogen-bonding parts.

3.9 MOLECULAR KINETICS

Molecules are in constant motion. At higher temperature the molecules move faster. If it were possible to get to the absolute zero temperature (−273.15 °C = −469.67 °F) motion would reach a minimum. There are three types of motion (Fig. 3.23). **Translation** involves the entire molecule moving

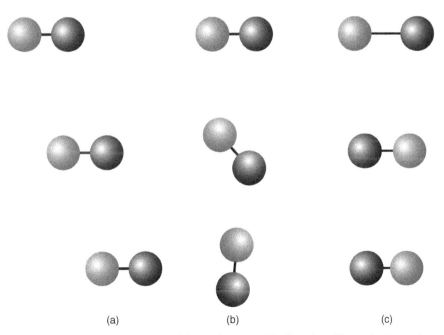

Figure 3.23 (a) Translation, (b) rotation, and (c) vibration. (See color insert.)

from one place to another. **Rotation** is the turning of the molecule about an axis. **Vibration** is the stretching or twisting of bonds so that the atoms in a molecule move with respect to one another. At higher temperature these motions become more energetic. Molecules that stick to one another by intermolecular forces at low temperature may fly apart when the temperature is raised. Solids may melt; liquids may vaporize. As the molecules move faster, they collide with one another harder, and more of these collisions have enough energy to break chemical bonds and cause reactions. As a result, nearly all chemical reactions go faster at higher temperature. Parts of molecules that are held in place by weak forces at low temperatures may be ripped from one another at higher temperatures. For this reason proteins, such as the amylase enzymes, unfold at high temperature and may lose some of their characteristics. They are said to **denature**.

3.10 CHEMICAL REACTIONS AND EQUATIONS

A chemical **reaction** is a process by which one or more substances (compounds or elements) change to other substances. The reaction occurs as a result of rearranging the atoms into new patterns characteristic of the new substances. The starting substances are called **reactants**; the final substances are called **products**. The atoms in the products have to be the same as those in the reactants because no atom is created or destroyed. A chemical reaction is often shown in the form of a **chemical equation**. The balanced chemical equation for the reaction of hydrogen with oxygen to give water is

$$2H_2 + O_2 \rightarrow 2H_2O$$

The arrow points from the reactants to the products and is read as "yields." The twos before the H_2 and the H_2O are called **coefficients**. Coefficients show how many of that type of molecule are used up or made by the reaction. The oxygen has a coefficient of one, which is not shown. The equation has four H atoms in the reactants, two hydrogen molecules with two hydrogen atoms each. It has four H atoms in the product, two water molecules with two hydrogen atoms each. There are two O atoms on the left, one oxygen molecule with two O atoms, and two on the right, two water molecules with one O atom each. An equation that has the same number of each type of atom on each side is **balanced**. Chemical equations are usually expected to be balanced. Sometimes, especially when we deal with a series of reactions in which products of each reaction become reactants in the next, it can be more understandable to present reactions in an unbalanced form. Various conventions are then used to designate reactants that did not originate with the previous step and products that do not participate in the next step.

3.11 MIXTURES

Beer, like almost every other material that we ordinarily encounter, has more than one compound or element. It is a **mixture**. The substances that make up the mixture are called **components**. The amount of each component making up the mixture is the **composition** of the mixture. A **solution** is a mixture that is well mixed all the way down to the molecular scale. An example of a solution is vinegar, a mixture of water and acetic acid. Any sample from any part of a solution (as long as the sample is much bigger than the molecules) has the same composition. A solution is also called a **homogeneous** [Greek: *homos*, same + *genos*, kind] mixture. Often one component of a solution makes up the majority of the solution. The dominant component is called the **solvent** and the smaller components are called **solutes**. In beer, the solvent is water; ethanol is one of the solutes. If the mixture does not mix well down to the molecular level it is a **heterogeneous** [Greek: *heteros*, different + *genos*, kind] mixture. A heterogeneous mixture has regions of different composition in different places. An example is oil and water. A less extreme example is milk, which is sugar-water with droplets of fat. The regions of a heterogeneous mixture are called **phases**. Milk has two phases, a sugar-water phase and a fat phase. The boundaries between phases in a heterogeneous mixture often scatter light, giving the mixture a cloudy appearance. Fresh bright beer in a bottle is a solution (or nearly so). Once the beer is opened, foam starts to form, so there is a liquid phase and a bubble phase. Stale beer is likely to become cloudy because of the formation of tiny solid particles, making the beer heterogeneous.

BIBLIOGRAPHY

Engdahl, A.; and Nelander, B. The Vibration Spectrum of H_2O_3. *Science*, **2002**, *295*(5554): 482–483. Isolation of dihydrogen trioxide.

Kotz, John, C.; Treichel, Paul M.; and Townsend, John R. *Chemistry and Chemical Reactivity*, 8th ed. Brooks/Cole, 2011. A well-presented standard textbook. General chemistry textbooks provide a wealth of detail and many tables of potentially useful information. Old editions are available cheap.

QUESTIONS

3.1. Indicate how many valence electrons the following atoms have: calcium, carbon, oxygen, chlorine.

3.2. What charge would be expected on ions of the following atoms: K, Al, Mg, F, S?

3.3. How many covalent bonds would we expect for each of the following atoms: carbon, nitrogen, hydrogen, oxygen?

3.4. Draw a structure of a molecule with all single bonds with two carbon atoms bound together and enough hydrogen atoms to satisfy the valency of the carbon atoms.

3.5. Give the shape about the central C atom for the following substances:

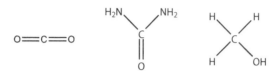

3.6. Give the underlying geometry and shape about the central C atom.

3.7. Give the underlying geometry and shape about the N atom.

3.8. Identify the intermolecular forces that would operate among molecules with the structure shown.

3.9. Identify the intermolecular forces that would operate among molecules with the structure shown.

3.10. Identify the intermolecular forces that would operate among molecules with the structure shown.

3.11. Give the formula for compounds made from (a) Ni^{2+} and OH^-, (b) Fe^{3+} and S^{2-}, and (c) K^+ and Br^-.

3.12. Based on information from Table 3.3, give formulas for compounds made with (a) H and Cl, (b) H and S, and (c) N and Cl.

3.13. For the following compounds, indicate whether they are held together by ionic bonds, covalent bonds, or both: H_2O, NaCl, CO_2, and $CaCO_3$.

3.14. Name and describe the three types of molecular motions.

3.15. Explain the difference between a chemical formula and a chemical equation. Give examples of each.

3.16. For each item, tell whether it is a mixture or a pure substance. For mixtures, identify the components and indicate whether the mixture is homogeneous or heterogeneous: Ethanol, vodka, blood, gasoline, air.

*__3.17.__ Draw the structures of OCl_2, NCl_3, and CCl_4, showing atoms, bonds, and unshared electrons. Give the underlying geometry and shape about the central atom. Tell whether the molecule is likely to be polar. Give the names for these compounds.

*__3.18.__ Give formulas for the following: potassium carbonate, sulfur trioxide, copper(I) nitrate, sodium hydrogen phosphate, and carbonic acid.

*__3.19.__ Give names for the following: P_2O_5, $FeSO_4$, $CaCl_2$, and NaOH.

APPENDIX TO CHAPTER 3

MEASUREMENT IN CHEMISTRY

This appendix about chemical measurements and calculations is essential for the practicing brewer, but is not necessary for a conceptual understanding of the chemistry of beer. It can be treated as supplementary material. You can skip all or part of this appendix and still understand the rest of the book.

NUMBERS

One difficulty in dealing with atoms and molecules is the scaling. Any property of a molecule is very small. For example, the mass of a molecule of ethanol is 0.0000000000000000000000764 gram. One gram of ethanol has 13090000000000000000000 molecules. Numbers like these are hard to read and write, and are prone to errors. When indicated, we will use exponential notation. Exponential notation is based on the idea that 10^n is 1 followed by n zeros. So 10^5 is 100,000. Conversely, 10^{-n} is $1/10^n$, which is what you get if you take 1.0 and move the decimal point n places to the left. So 10^{-5} is 0.00001. We can express any number as a number of moderate size times a power of 10. The mass of a molecule of ethanol would be written 7.64×10^{-23} gram and the number of molecules in a gram of ethanol would be written 1.309×10^{22}. Scientific calculators provide for the input and output of numbers in exponential notation.

Most numbers used in chemistry have units. That is, they are in grams or milliliters or something. A number without its units is naked. Unless the

---

The Chemistry of Beer: The Science in the Suds, First Edition. Roger Barth.
© 2013 John Wiley & Sons, Inc. Published 2013 by John Wiley & Sons, Inc.

number is the count of something (class size is 62), or a pure fraction (male to female ratio is 1.05), it needs a unit. There is a man-made crater on Mars where, in 1999, a 125 million dollar orbiter crashed because two engineering teams working on the project were using different units, and they were not in the habit of including the units with their numbers.

INTERNATIONAL SYSTEM

Chemistry depends on measurements. Because chemistry is an international enterprise, chemistry measurements are almost always reported in the metric system. The metric system, which is based on an internationally agreed set of standards, is called the International System and is symbolized SI. Use of metric measurements avoids some annoying sources of confusion. For example, the volume called one gallon is 20% larger in Britain and Ireland than in the United States. The volume called a barrel depends on what it is a barrel of. A U.S. beer barrel is 31 U.S. gallons (117.3 liters), but a British beer barrel is 36 Imperial gallons, or about 43 U.S. gallons (163.7 liters). A U.S. oil barrel is 42 U.S. gallons (159.0 liters). The use of the metric system allows clearer, more error-free communication.

The metric system provides for seven basic dimensions of measurement: length, time, mass, electric current, temperature, amount of substance, and luminous intensity. Anything else that one may want to measure is based on these seven. For example, speed is distance (length) divided by time.

SI Base Units

There is a distinction between the **dimension**, which is the type of quantity that we measure, like length or time, and the **units**, which are the standard quantities used to express measurements, like meters or seconds. The SI base units are the fundamental units for measuring the seven dimensions. For example, the SI base unit for length is the meter. When written with a number, the meter is given the symbol m.

SI BASE UNITS

Dimension	Unit	Symbol
Length	meter	m
Mass	kilogram	kg
Time	second	s
Electric current	ampere	A
Temperature	kelvin	K
Amount of substance	mole	mol
Luminous intensity	candela	cd

Of the basic dimensions, length, mass, and time need no explanation. Electric current is the speed of flow of electric charge. Charge is measured by a derived unit called the coulomb (C) equal to one ampere times one second. The unit of temperature is the kelvin (the degree sign is not used). A kelvin is the same size as the degree Celsius (centigrade), which comes to 1.8 Fahrenheit degrees. The kelvin temperature scale is an absolute scale, meaning that it starts at the minimum possible temperature, not some arbitrarily chosen zero point. The zero of the kelvin scale is at −273.15 °C (−459.67 °F). Zero kelvin is the lowest possible temperature; thermal energy is at a minimum. The amount of substance is a measure of the number of particles (molecules or atoms). It is a complex concept that plays a key role in chemistry. We will discuss amount of substance in its own section. Luminous intensity does not seem to be used by most chemists.

There is an older version of the metric system called the cgs (centimeter–gram–second) system in which the base unit for length is the centimeter and that for mass is the gram. The main survivors from this system are a unit for energy called the erg ($g \cdot cm^2/s^2$) and one for force called the dyne ($g \cdot cm/s^2$). One erg is 10^{-7} joule and one dyne is 10^{-5} newton (1 N = 1 $kg \cdot m/s^2$). Surface tension is often expressed in cgs units (dyne/cm).

Sometimes, for convenience, we use prefixes to produce a multiple of a unit. For example, the prefix kilo- refers to 1000 of something, so a kilometer (km) is 1000 meters. Prefixes can also provide smaller units, so a millimeter (mm) is one-thousandth of a meter (0.001 m). Some of the SI prefixes are shown below.

SI PREFIXES

10^3	kilo	k	10^{-3}	milli	m
10^6	mega	M	10^{-6}	micro	μ
10^9	giga	G	10^{-9}	nano	n
10^{12}	terra	T	10^{-12}	pico	p
10^{15}	peta	P	10^{-15}	femto	f
10^{18}	exa	E	10^{-18}	atto	a

Some prefixes that are used in certain contexts, but are not officially accepted, are hecto (h, 100) and centi (c, 0.01).

Derived Units

Base units can be combined to give different dimensions of measurement. To measure volume, we can consider a cubic box one meter on an edge. The volume of a box is the length times the width times the height, so the 1 m cube would have a volume of 1 m × 1 m × 1 m = 1 m^3. The cubic meter (m^3) is the derived unit for volume. Many derived units have their own names. An example is the unit for energy, $kg \cdot m^2/s^2$, which is called a joule (J).

We will discuss a few important dimensions with derived units.

Volume

The SI unit for volume is the cubic meter. Regrettably, the cubic meter does not have its own name. The cubic meter is a very large volume; one cubic meter of water weighs 1000 kilograms (1 Mg, called a tonne, about 2200 pounds). A commonly used unit for volume is the liter (L). A liter is 0.001 cubic meter. Beer is often measured in hectoliters (hL = 100 L). One hectoliter is equivalent to 0.852 U.S. beer barrels.

Energy

The SI unit of energy is the joule (J), equivalent to kg·m^2/s^2. The energy needed to raise one kilogram by one meter at the surface of the earth is about 10 J. Energy is sometimes measured in calories (the energy needed to raise the temperature of one gram of water by one kelvin). A calorie is equivalent to 4.184 J. The energy provided by food is sometimes measured in Calories (capital C). A Calorie is 1000 calories. Another energy unit used by engineers is the British thermal unit (BTU). One BTU is equivalent to 1055 J. When heating or cooling power is specified in BTU, the intended unit is BTU per hour. The electric company measures electrical energy in kilowatt hours (kWh). A kWh is equivalent to 3.6×10^6 J or 3.6 megajoules (MJ).

Force

The SI unit of force is the newton (N). A newton is 1 kg·m/s^2. One newton is equivalent to about one-fifth of a pound. Force is sometimes confused with mass, because the English system does not have a consistent unit for mass. The force that an object exerts as a result of gravity is called its weight. An object in orbit will be weightless, but it will still have mass; that is, it will require the application of force to speed it up or slow it down. We usually determine the mass of an object by comparing its weight with that of a known mass. For this reason, determining the mass of something is called weighing (we do not use "mass" as a verb in this book).

Pressure

Pressure is force divided by the area on which the force acts. The smaller the area; the higher the pressure. This recognizes that a certain force applied with the palm of one's hand can have a different effect than the same force applied with an ice pick. The SI unit of pressure is the pascal (Pa), a newton per square meter, equivalent to kg/(m·s^2). A pascal is a small unit of pressure; the pressure of the atmosphere at sea level is over 10^5 Pa.

A common unit of pressure is the pound per square inch (psi). A psi is equivalent to 6894 Pa. Pressure is also measured in atmospheres (atm). One standard atmosphere is 101325 Pa, the usual air pressure at sea level.

Amount of Substance (Moles)

Sometimes we need to keep track of the number of atoms or molecules in a sample. We can't go strictly by mass; different atoms and molecules have different masses. One gram of water has over 2½ times as many molecules as one gram of ethanol because ethanol molecules are much more massive than those of water. It is not possible to count atoms and molecules because they are invisibly small and uncountably numerous. To understand how chemists keep track of particles, it is helpful first to consider how builders keep track of nails.

A building project uses many more nails than anyone wants to count, so instead, nails are weighed. To plan a project, the builder consults a table that gives the weight of 1000 nails. For example, 1000 1½ inch tinned nails weighs 4 pounds. Based on this information, the builder knows how to count nails by weighing. It should be clear that this method will only work if all the nails are about the same mass.

To put this into a chemical perspective, a single one of these nails contains about 19,600,000,000,000,000,000,000 iron atoms. Clearly we can't deal with atoms in thousands; the weight of a thousand atoms would not register on the most sensitive scale. The standard number of units—that is, atoms, molecules, or ions—used by chemists is 6.022×10^{23}, which is 6.022 multiplied by 1 followed by 23 zeros. The reason for this clumsy-looking number is that it is about the number of atoms in one gram of hydrogen, the lightest of the atoms. An amount of stuff with this many particles is called a **mole** of the stuff. In chemical jargon, we say that the **molar mass** (also called *atomic weight*) of hydrogen atoms is one gram per mole. The mass of a mole of something is called its molar mass or molecular weight (in the special case of the atoms of an element, it is sometimes called the atomic weight). The molar masses of atoms of the elements are determined by the numbers of protons and neutrons in the atoms. Different elements have different molar masses, because their atoms weigh different amounts. The molar masses of the elements are listed on the periodic table facing the opening page of Chapter 3.

Rather than think of H_2O as a formula representing a single water molecule weighing 3.0×10^{-23} gram, we can think of it as representing a mole of water molecules weighing 18 grams. The formula tells us that each mole of water has two moles of hydrogen atoms (1 gram each) and one mole of oxygen atoms (16 grams each). We can think of chemical equations in terms of moles. The equation $H_2 + \frac{1}{2}O_2 \rightarrow H_2O$ makes sense if we read it as one mole of H_2 reacts with half a mole of O_2 and yields one mole of H_2O.

CONVERSION TABLES

To use the tables below to convert from the unit in the first column to that in the first row, multiply by the number in the indicated box. For example, to convert 50 grams to ounces, find grams in the *from* column and ounces in the *to* row. The number in the box where these meet is 0.03527, so the number of ounces in 50 grams is 50 g × 0.03527 oz/g = 1.764 oz.

Mass Conversions

from \ to	gram	kilogram	pound	ounce	U.S. ton	tonne
gram	1	0.001	0.002205	0.03527	1.102×10^{-6}	1×10^{-6}
kilogram	1000	1	2.205	35.27	0.001102	0.001
pound	453.59	0.45359	1	16	0.0005	4.536×10^{-4}
ounce	28.35	0.02835	0.0625	1	3.215×10^{-5}	2.835×10^{-5}
U.S. ton	907.18	907.2	2000	32000	1	0.9072
tonne	1×10^6	1000	2205	35274	1.102	1

Volume Conversions

from \ to	milliliter	liter	hectoliter	U.S. fl oz	U.S. gal	U.S. beer bbl
mL	1	0.001	1×10^{-5}	0.03381	2.642×10^{-4}	8.522×10^{-6}
L	1000	1	0.01	33.81	0.2642	0.008522
hL	1×10^5	100	1	3381	26.42	0.8522
U.S. fl oz	29.57	0.2957	2.957×10^{-4}	1	0.007812	2.520×10^{-4}
U.S. gal	3785	3.785	0.03785	128	1	0.03226
Imp gal	4546	4.546	0.04546	153.7	1.201	0.03874
U.S. bbl	117348	117.35	1.1735	3968	31	1
U.K. bbl	163659	163.66	1.6366	5534	43.23	1.395

Volume to Mass for Water at 20 °C

from \ to	gram	kilogram	ounce	pound
milliliter	0.9982	9.982×10^{-4}	0.03521	0.002201
liter	998.2	0.9982	35.21	2.201
hectoliter	99821	99.82	3521	220.1
fl oz	29.52	0.02952	1.0412	0.06507
U.S. gal	3778	3.778	133.26	8.329
U.S. beer bbl	117123	117.12	4131	258.2

(continued)

Pressure Conversions

from \ to	kilopascal	atmosphere	Torr (mm Hg)	inch Hg	psia
kPa	1	0.009869	7.501	0.2953	0.14303
atm	101.325	1	760	29.29	14.695
Torr	0.13332	0.0013158	1	0.03937	0.19264
in Hg	3.386	0.03342	25.4	1	0.4893
psia	6.895	0.06805	51.91	2.044	1

Celsius to Fahrenheit Conversion

°C	°F	°C	°F	°C	°F	°C	°F
0	32.0	26	78.8	52	125.6	78	172.4
2	35.6	28	82.4	54	129.2	80	176.0
4	39.2	30	86.0	56	132.8	82	176.6
6	42.8	32	89.6	58	136.4	84	183.2
8	46.4	34	93.2	60	140.0	86	186.8
10	50.0	36	96.8	62	143.6	88	190.4
12	53.6	38	100.4	64	147.2	90	194.0
14	52.7	40	104.0	66	150.8	92	197.6
16	60.8	42	107.6	68	154.4	94	201.2
18	64.4	44	111.2	70	158.0	96	204.8
20	68.0	46	114.8	72	161.6	98	208.4
22	71.6	48	118.4	74	165.2	100	212.0
24	75.5	50	122.0	76	168.8		

To calculate the molar mass of a compound, we add together the molar masses of the atoms in it. So for a mole of ethanol, C_2H_6O, we have

- 2 moles of carbon with a molar mass of 12.011 grams per mole contributes 24.022 g/mol,
- 6 moles of hydrogen at 1.008 g/mol contributes 6.048 g/mol,
- 1 mole of oxygen at 15.999 g/mol. The net molar mass is 24.022 g/mol + 6.048 g/mol + 15.999 g/mol = 46.069 g/mol.

This calculation is shown below:

Element	Number	Molar Mass	Total
Carbon	2	12.011 g/mol	24.022 g/mol
Hydrogen	6	1.008	6.048
Oxygen	1	15.999	15.999
Total			46.069 g/mol

To calculate the amount (number of moles) of a substance from the mass, we divide the mass by the molar mass. For example, the amount of ethanol in 10 g of ethanol is 10 g ÷ 46.069 g/mol = 0.217 mol. The mass of water that would have the same number of molecules would be 0.217 mol × 18 g/mol = 3.91 g (18 g/mol is the molar mass of water). The three equations (they are algebraic variations of a single equation) below summarize calculations involving mass and moles.

$$\text{moles} = \frac{\text{mass}}{\text{molar mass}} \qquad \text{molar mass} = \frac{\text{mass}}{\text{moles}}$$

$$\text{mass} = \text{moles} \times \text{molar mass}$$

MASS RELATIONSHIPS IN COMPOUNDS

Suppose we want 6.3 grams of calcium ion in a batch of beer. Calcium itself is a reactive metal that is dangerous to handle. We don't want to put it into our beer. Instead, we might prefer to start with calcium chloride, $CaCl_2$. But when we put $CaCl_2$ on the scale, we are weighing the chloride with the calcium. We will need more than 6.3 grams because not everything we are weighing out is calcium ion. The way we attack this problem is to remember that the formula, $CaCl_2$, tells us that each mole of compound contains one mole of calcium and two moles of chlorine.

- First we calculate the moles in 6.3 grams of calcium, whose molar mass from the periodic table is 40.078 g/mol.
- The number of moles of calcium comes to 6.3 g ÷ 40.078 g/mol = 0.157 mol.
- This is equal to the moles of $CaCl_2$, whose molar mass is 40.078 g/mol + 2 × 35.453 g/mol = 110.98 g/mol.
- The mass of $CaCl_2$ required is 0.157 mol × 110.98 g/mol = 17.4 grams.

We can do this type of calculation in reverse to determine how much of an element is in a known amount of compound.

What is the mass of oxygen in 74 grams of acetic acid, $C_2H_4O_2$?
- The molar mass of acetic acid is 60 g/mol and that of oxygen atoms is 16 g/mol.
- We know the mass and molar mass of acetic acid, so we can calculate the moles: 74 g ÷ 60 g/mol = 1.23 mol.
- The formula of acetic acid tells us that for each mole of acetic acid, there are two moles of oxygen, so there are 2 × 1.23 mol = 2.46 moles of oxygen.
- The mass of oxygen is 2.46 mol × 16 g/mol = 39.5 grams.

COMPOSITION OF MIXTURES

Beer and its ingredients are mixtures. To describe a mixture we need to identify the components and to specify how much of each is present. This specification is called the **composition** of the mixture.

Mass Percent

Suppose we prepare a mixture with 5 grams of ethanol (drinking alcohol) and 95 grams of water. Experience tells us that the properties of the mixture would be the same if we had doubled the recipe and used 10 grams of ethanol and 190 grams of water. One way to specify the composition of this mixture would be to divide the mass of each component by the total mass of the mixture. For 5 grams of ethanol and 95 grams of water, the total mass is 100 grams, so the mass fraction of ethanol in the mixture is 5 g ÷ 100 g = 0.05 and that of water is 95 g ÷ 100 g = 0.95. These same fractions would result from the doubled recipe of 10 g of ethanol and 190 g of water. Often we multiply the mass fraction by 100% to give a percentage: 5% ethanol and 95% water. The mass percent is sometimes symbolized by % w/w (for percent weight by weight), to distinguish it from other percentage measures of composition. In chemistry, an unqualified percent refers to mass percent. For substances present in trace amounts, we might multiply the mass fraction by 10^6 ppm to give parts per million (ppm), by 10^9 ppb to give parts per billion, (ppb), or even by 10^{12} ppt to give parts per trillion (ppt).

$$\%w/w = \frac{\text{mass component}}{\text{total mass}} \times 100\%$$

Weight–Volume Percent

The mass (in grams) of a substance in 100 milliliters of a mixture is called the weight–volume percent. To calculate weight–volume percent, take the mass of solute in grams, divide by the volume in milliliters, then multiply by 100%. The equation for this operation is

$$\%w/v = \frac{\text{mass component (g)}}{\text{total volume (mL)}} \times 100\%$$

If we dissolve 5 grams of glucose in enough water to give 250 mL of solution, the weight–volume percent glucose would be (5 grams)/(250 mL) × 100% = 2% w/v.

Volume Percent

The volume fraction is the volume of the component divided by the volume of the mixture. The volume percent is the volume fraction times 100%:

$$\%v/v = \frac{\text{volume component (mL)}}{\text{total volume (mL)}} \times 100\%$$

If we have 355 mL of beer that contains 15 mL of ethanol, the volume fraction would be 15 mL/355 mL = 0.042. This comes to 4.2% v/v, or 4.2% ABV (alcohol by volume). Volume fraction is most often used when the component of interest is a liquid or a gas. A complication of volume fraction is that the total volume of the solution is not always exactly equal to the sum of the volumes of the components. As a result, the sum of the volume percents of all components may not come to exactly 100%.

Molar Concentration

If we need to deal with mixtures on the basis of the number of particles, we would keep track of the composition by moles. The standard measure of composition by moles is the number of moles in each liter of mixture. This measure is the **molar concentration**, symbolized c, also called the **molarity** of the mixture. The molar concentration can be calculated from the number of moles of a substance divided by the volume of solution, in liters:

$$c = \frac{\text{moles solute}}{\text{total volume (L)}}$$

Suppose we take 5 grams of glucose ($C_6H_{12}O_6$ molar mass = 180.156 g/mol) and dissolve it in enough water to give 250 milliliters (0.25 L). The amount of glucose is (5 g)/180.156 g/mol = 0.02775 mol. The molar concentration is (0.02775 mol)/(0.25 L) = 0.111 mol/L.

Composition Formulas

Mass fraction: $\quad\text{mass fraction} = \dfrac{\text{mass component}}{\text{total mass}}$

Molar concentration: $\quad c = \dfrac{\text{moles solute}}{\text{total volume (L)}}$

Weight–volume percent: $\quad \%w/v = \dfrac{\text{mass component (g)}}{\text{total volume (mL)}} \times 100\%$

Volume fraction: $\quad \text{volume fraction} = \dfrac{\text{volume component}}{\text{total volume}}$

Examples

How many moles are in 38.5 g of maltose, $C_{12}H_{23}O_{11}$?

Answer: The molar mass of maltose is 12×12 g/mol $+ 23 \times 1$ g/mol $+ 11 \times 16$ g/mol $= 343$ g/mol. The number of moles (n) is $n = $ mass/MM $= 38.5$ g / 343 g/mol $= 0.112$ mol.

Calculate the mass of 55.5 moles of water, H_2O.

Answer: The molar mass is 2×1 g/mol $+ 1 \times 16$ g/mol $= 18$ g/mol. The mass is the number of moles times the molar mass: Mass $= n \times$ MM $= 55.5$ mol $\times 18$ g/mol $= 999$ g.

Calculate the mass of glucose (MM = 180 g/mol) that has the same number of molecules as 12 g of ethanol, C_2H_6O.

Answer: Same number of molecules means the same number of moles. The molar mass of ethanol is 2×12 g/mol $+ 6 \times 1$ g/mol $+ 1 \times 16$ g/mol $= 46$ g/mol. The number of moles of ethanol is 12 g / 46 g/mol $= 0.26$ mol. The number of moles of glucose is the same, 0.26 mol. The mass of 0.26 mol of glucose is mass $= n \times$ MM $= 0.26$ mol $\times 180$ g/mol $= 47$ g.

What mass of potassium sulfate (K_2SO_4) contains 3.0 g of potassium?

Answer: The molar mass of K_2SO_4 is 174.3 g/mol and that of potassium is 39.098 g/mol. The number of moles of K is 3.0 g ÷ 39.098 g/mol $= 0.0767$ mol. One mole of K_2SO_4 has two moles of K, so each mole of K goes with half a mole of K_2SO_4. The number of moles of K_2SO_4 is 0.0767 mol $\times$ ½ $= 0.0384$ mol. The mass of K_2SO_4 is 0.0384 mol $\times$ 174.3 g/mol $= 6.7$ grams.

What mass of maltose and what mass of water would need to be combined to make 350 grams of an 11% by mass maltose solution?

Answer: Eleven percent corresponds to a mass fraction of 0.11. The mass of maltose would be 0.11×350 g $= 38.5$ g of maltose. The rest is water: $350 - 38.5$ g $= 311.5$ g.

What volume of a 0.15 mol/L solution of acetic acid would contain 0.0056 mol of acetic acid.

Answer: Number of moles = volume × molar concentration: $n = V \times c$. $V = n/c$. $V = 0.0056$ mol / 0.15 mol/L $= 0.037$ L or 37 milliliters.

BIBLIOGRAPHY

International Union of Pure and Applied Chemistry. *Quantities, Units, and Symbols in Physical Chemistry*, 3rd ed. RSC, 2007. Called the Gold Book. All the rules, some of which are seldom obeyed.

QUESTIONS

3A.1. Convert 12 kilometers to meters.

3A.2. Calories are units for what quantity?

3A.3. Calculate the molar mass of lactic acid, $C_3H_6O_3$.

3A.4. Calculate the number of moles in 17 grams of lactic acid.

3A.5. If 17 grams of lactic acid is dissolved in enough water to give 300 mL of solution, calculate the molar concentration of lactic acid in the solution.

3A.6. If 17 grams of lactic acid is dissolved in 300 grams of water, calculate the mass fraction and mass percent lactic acid.

3A.7. What is the SI unit for pressure?

3A.8. If 50 mL of phosphoric acid is dissolved in enough water to give 350 mL of solution, calculate the volume fraction and the volume percent phosphoric acid (H_3PO_4) in the solution.

3A.9. A mixture of solids to be used to make glass is prepared. The mixture contains 75 pounds of limestone ($CaCO_3$), 375 pounds of sand (SiO_2), and 50 pounds of soda ash (Na_2CO_3). Calculate the weight percent of each ingredient in the mixture.

3A.10. Calculate the mass in pounds of 5.5 U.S. gallons of water.

3A.11. Calculate the mass in ounces of 22 kg.

3A.12. Calculate the pressure in psia of 720 Torr.

3A.13. A recipe calls for mash-in at 65 °C. What is this to the nearest degree in °F?

*3A.14. What mass of iron would have the same number of atoms as 15 g of carbon? (Use the molar masses from the periodic table at the beginning of Chapter 3.)

*3A.15. Calculate the mass of magnesium iodide, MgI_2, that contains 4.2 grams of iodide ion.

*3A.16. What mass of an 11% solution of maltose ($C_{12}H_{22}O_{11}$) would have 68 grams of maltose?

*3A.17. What volume of a 0.25 mol/L solution of sodium chloride would contain 5.7 moles of sodium chloride?

Mediterranean Sea, viewed from Rosh HaNikra, Israel. Photo by David Barth.

CHAPTER 4

WATER

Beer is over 90% water; the human body is about 60% water. Because of its unique properties, water forms the matrix for all life on earth. Water is the only common substance that occurs in the solid, liquid, and gas phases at ordinary temperatures.

It takes between 3.5 and 6 gallons of water to make one gallon of beer. In addition to the water in the beer itself, water is used for cleaning and to move heat in and out during heating and cooling. Except in controlled laboratory situations, water always contains additional substances. If we don't like these substances, we regard them as **pollutants** and try to take them out. Beer cannot be made from perfectly pure water; the processes that turn barley starch to alcohol and carbon dioxide depend on the presence of traces of metal ions like calcium ion (Ca^{2+}), copper(II) ion (Cu^{2+}), and zinc ion (Zn^{2+}). In addition, the flavor of beer is influenced by certain trace components, such as bicarbonate (HCO_3^-) and sulfate (SO_4^{2-}) ions. Brewers sometimes call water that is treated and ready for use to make beer **brewing liquor**. Brewing liquor is not usually the same quality of water as that used to clean out the equipment or to provide heating or cooling.

4.1 THE WATER MOLECULE

The water molecule (Fig. 4.1) has an oxygen atom with two bonds, each to a hydrogen atom. The formula is H_2O. The oxygen atom has, in addition to the

---

The Chemistry of Beer: The Science in the Suds, First Edition. Roger Barth.
© 2013 John Wiley & Sons, Inc. Published 2013 by John Wiley & Sons, Inc.

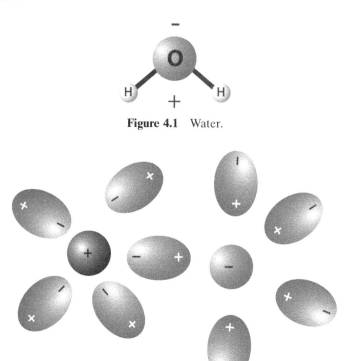

Figure 4.1 Water.

Figure 4.2 Hydrated ions.

two shared electron pairs forming the bonds, two unshared pairs. The four electron groups arrange themselves to be as far from one another as possible, giving a tetrahedral geometry whose ideal angle is about 109.5 degrees. In water the actual hydrogen–oxygen–hydrogen angle is about 104.5 degrees. This is far from a straight line (180 degrees), so water is bent. Because of this bend, water has a hydrogen end and an oxygen end.

Oxygen attracts the shared electrons to a greater extent than hydrogen, so the oxygen end of water has some negative charge, and the hydrogen end has some positive charge, making water a polar molecule. Because the water molecule has positively and negatively charged ends, it can arrange itself to bring the end of opposite charge near any ions or polar molecules. Molecules or ions that are surrounded by water molecules in this way are said to be **hydrated** (Fig. 4.2). Hydration gives a strong force of attraction allowing many ionic and polar molecules to dissolve in water.

The water molecule has two O–H bonds, each of which can form hydrogen bonds to other water molecules or to molecules of other hydrogen-bonding substances. It takes more energy to pull water molecules away from one another than most substances because of the strong forces of attraction between them. As a result, it takes more energy to melt or vaporize water than most materials. It takes more energy to warm a given mass of water by one degree than nearly anything else. This makes water an ideal material to carry heat. It also means that the heating and boiling operations in brewing require a large energy input. Another result of the strong force between water molecules is that nonpolar molecules or parts of molecules are squeezed out from between the water molecules so the water molecules can interact with one another. Molecules that have a polar or charged end attached to a nonpolar chain tend to clump with the nonpolar parts together and the polar parts pointing into the water. This behavior is important in the formation of foam and haze in beer and of biological structures, including the membranes of yeast cells.

4.2 ACIDS AND BASES

One of the important processes in a water solution is the transfer of a hydrogen ion (H^+) from one molecule or ion to another. As we will see, these processes can affect the electrical charges of species in contact with the solution, which can influence their behavior.

Water molecules can accept hydrogen ions from other molecules or ions, or they can provide hydrogen ions. A hydrogen ion is a hydrogen atom that has left its electron behind. Because a hydrogen atom only starts with one electron, when this is lost the hydrogen ion is nothing but a hydrogen nucleus, which consists of a single proton. A substance that provides a hydrogen ion is called an **acid**. A substance that accepts a hydrogen ion is called a **base**. Water can be an acid or a base. An example of the action of an acid in water is

$$HCl + H_2O \rightarrow Cl^- + H_3O^+$$

HCl (hydrochloric acid) provides a hydrogen ion to water, giving H_3O^+ (the **hydronium ion**) and the chloride ion. As shown in Figure 4.3, the bond between the hydrogen and chlorine breaks, leaving the shared electrons on the chlorine. The hydrogen ion forms a new bond by sharing an electron pair on the water molecule, forming hydronium ion. HCl is the acid and H_2O is the base. The hydronium ion is the signature of an acid in water. The concentration of H_3O^+ is a measure of the solution's acidity.

A substance that accepts a hydrogen ion is a **base**. An example of the action of a base in water is

$$NH_3 + H_2O \rightarrow NH_4^+ + OH^-$$

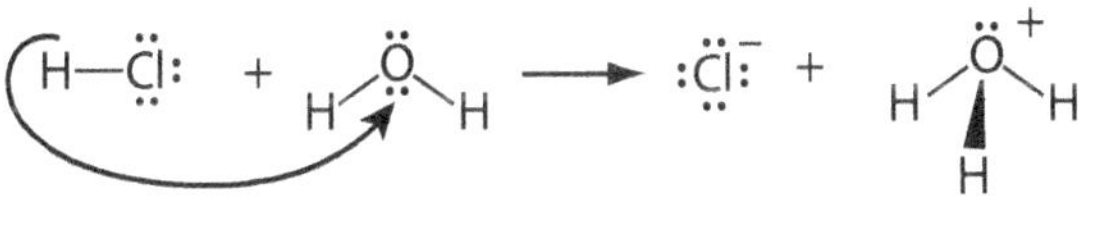

Figure 4.3 Acid–base reaction.

NH$_3$ (ammonia) accepts a hydrogen ion from water, leaving behind OH$^-$ (the hydroxide ion) and NH$_4^+$ (the ammonium ion). In this reaction, NH$_3$ is the base and H$_2$O is the acid. The hydroxide ion is the signature of a base in water. We can see that in the reaction with hydrochloric acid, water accepts the hydrogen ion that HCl provides; in the reaction with ammonia, water provides the hydrogen ion that NH$_3$ accepts. So with HCl, water acts as the base, but with NH$_3$, water acts as the acid; water can act either as an acid or as a base. If water can act as an acid or as a base, can one water molecule provide a hydrogen ion to another water molecule? It can indeed; a tiny fraction of the water undergoes a reaction, H$_2$O + H$_2$O → H$_3$O$^+$ + OH$^-$, where the water molecule on the left has accepted a hydrogen ion from the water molecule on the right.

The concentrations of hydroxide ion and hydronium ion are inversely related to one another; when one goes up, the other goes down. It turns out that, in water, the molar concentration of the hydronium ion times that of the hydroxide ion is a constant: Conc (H$_3$O$^+$) × Conc (OH$^-$) = K_w, where K_w, a temperature-dependent number called the ion product of water, has a value at room temperature of about 10^{-14}. This equation tells us that anything that raises the concentration of H$_3$O$^+$ must also lower the concentration of OH$^-$ to keep the value of K_w constant. Water with equal concentrations of H$_3$O$^+$ and OH$^-$ is considered **neutral**. Water with more H$_3$O$^+$ must have less OH$^-$. Such water is **acidic**. Water with less H$_3$O$^+$ and more OH$^-$ is **basic**, or **alkaline**.

Strong and Weak Acids and Bases

A strong acid easily donates its hydrogen ion. In water, each strong acid molecule donates a hydrogen ion to water, giving hydronium ion, H$_3$O$^+$. A weak acid donates its hydrogen ion more reluctantly. In water, only a fraction of the weak acid molecules donate a hydrogen ion to give hydronium ion. The same distinction applies to bases. Every molecule of a strong base accepts a hydrogen ion from water, leaving behind a hydroxide ion, OH$^-$. There are many weak acids and bases, but only a few strong ones. Table 4.1 gives some examples of strong and weak acids and bases.

Le Châtelier's Principle

The behavior of H$_3$O$^+$ and OH$^-$ is an example of a general rule called Le Châtelier's principle. This rule states that a reaction tends to move in the direction to use up any reactant or product that is present in excess, and to make more of a reactant or product that is in short supply. When the concentrations of reactants and products are in balance, the reaction is at equilibrium. All reactions have a spontaneous tendency to move toward equilibrium. For the reaction

$$H_2O \rightarrow H_3O^+ + OH^-$$

TABLE 4.1 Common Acids and Bases

Strong Acids		Weak Acids	
Acid	Formula	Acid	Formula
Sulfuric	H_2SO_4	Acetic	CH_3CO_2H
Hydrochloric	HCl	Butyric	$C_3H_7CO_2H$
Nitric	HNO_3	Lactic	$CH_3CHOHCO_2H$
Perchloric	$HClO_4$	Phosphoric	H_3PO_4

Strong Bases		Weak Bases	
Base	Formula	Base	Formula
Sodium hydroxide	NaOH	Ammonia	NH_3
Potassium hydroxide	KOH	Bicarbonate ion	HCO_3^-
		Methylamine	CH_3NH_2

if we do something to increase hydronium ion (H_3O^+) the reaction will try to use some of it up by going toward reactants, and OH^- will decrease. If we do something to decrease hydroxide ion (OH^-), the reaction will go farther toward products to replace some of the OH^-, and H_3O^+ will increase.

4.3 pH

We follow the acidity/basicity of water with the **pH**. On the pH scale, a pH of 7 represents a neutral solution. When the solution becomes more acidic, the pH goes down. A solution whose pH is less than 7 is acidic. If the pH is greater than 7, it is basic. When the pH goes down by one unit, the concentration of H_3O^+ goes up by a factor of 10, and the concentration of OH^- goes down by a factor of 10. Such a scale is called a logarithmic scale. The pH of beer is generally about 4 (Table 4.2). The lowest possible (though not practically attainable) pH in water is –1.7. This would require every water molecule to grab a hydrogen ion and form H_3O^+. The highest possible pH in water, requiring that every water molecule give up a hydrogen ion to form OH^-, is 15.7. Devices to measure pH generally work in the range from pH 1 to 14 (Fig. 4.4). The pH is a key issue in all chemistry involving water, including beer chemistry.

To highlight the importance of pH in brewing, we will consider the role of proteins. Proteins are very long molecules that are involved in the fermentation process, in beer foam quality, and in clarity or cloudiness of beer. Proteins often have many sites that can act as acids or as bases. When the solution is acidic (low pH), some of the basic sites on a protein will pick up positively charged hydrogen ions from the hydronium ions in the solution. These sites will become positively charged. If the solution is basic (or less acidic), some of the acidic sites will give up hydrogen ions to the solution and take a negative charge. These situations are illustrated in Figure 4.5. The behavior of the

74 WATER

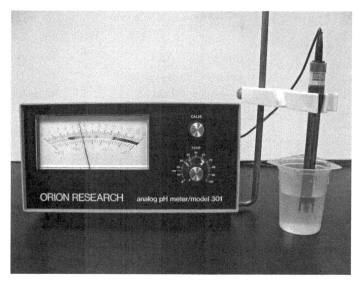

Figure 4.4   pH meter. Photo by Roger Barth.

TABLE 4.2   pH of Common Materials

Material	pH
Lemon juice	2
Beer	4
Milk	6.7
Blood	7.2
Baking soda	9
Oven cleaner	13

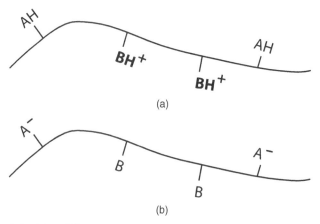

Figure 4.5   (a) Acidic solution—low pH and (b) basic solution—high pH.

protein will change depending on the distribution of positive and negative charges, and hence on the pH of the solutions. The charge on the molecule changes with pH, getting more positive at low pH and more negative at high pH. The pH at which the positive and negative charges balance, making the molecules electrically neutral (on average), is called the **isoelectric point**. Molecules have a net positive charge at a pH lower than the isoelectric point and a net negative charge at a pH higher than the isoelectric point.

---

pH: A CLOSER LOOK

The pH scale is based on the molar concentration of hydronium ion, H_3O^+. A pH of 1 represents a hydronium ion concentration of 10^{-1} mol/liter. A pH of 7 represents a concentration of 10^{-7} mol/liter. To deal with hydronium ion concentrations that do not fall out as even powers of 10, we use logarithms.

$$pH = -\log[H_3O^+]$$

where $[H_3O^+]$ symbolizes "the molar concentration of hydronium ion." So if the molar concentration of hydronium ion is 5×10^{-5} mol/L, the pH is $-\log(5 \times 10^{-5}) = 4.3$.

The molar concentration of hydronium ion is related to that of hydroxide ion at 25 °C by the equation $[H_3O^+][OH^-] = 10^{-14}$. In a neutral solution the hydronium and hydroxide concentrations are equal. This equation tells us that in a neutral solution they must each be 10^{-7} mol/L. Hence a neutral solution has a pH of 7. When the pH is lower than 7 the concentration of hydronium ion is higher and that of hydroxide ion is lower.

The pH governs the extent to which an acid or base exists with or without a hydrogen ion. Lactic acid, which gives a tart flavor to some sour beers like lambic, can exist in either of two forms: one with the hydrogen ion—$CH_3CH(OH)COOH$ (lactic acid); and the other without the hydrogen ion—$CH_3CH(OH)COO^-$ (lactate ion). When the pH is 3.9 (a value called the pK of lactic acid) the two forms are at equal concentration. At higher pH the solution is more basic; that is, it is deficient in hydrogen ions so the concentration of lactate ion becomes higher and that of lactic acid becomes lower. More of the molecules exist in the negatively charged form. At low pH the uncharged lactic acid form dominates at the expense of the lactate ion. The pK is different for different acids and bases; hence the pH at which the concentrations of the two forms are equal depends on the identity of the acid or base.

Acidic groups on proteins and other complex molecules take a negative charge at high pH. Basic groups take a positive charge at low pH. This can affect the functioning of the protein. The pH at which the molecule is electrically neutral is called the **isoelectric point**.

TABLE 4.3 Effect of Ions on Beer

Ion	Formula	Effect
Bicarbonate	HCO_3^-	Increase pH
Calcium	Ca^{2+}	Decrease pH
Chloride	Cl^-	Sweetness, fullness, balance sulfate
Iron(II)	Fe^{2+}	Metallic, astringent
Hydronium	H_3O^-	Decrease pH, enhance bitterness
Magnesium	Mg^{2+}	Required by yeast
Sodium	Na^+	Sweet, sour at higher levels
Sulfate	SO_4^{2-}	Dryness, astringency, enhances bitterness

4.4 IONS AND BEER

The ions in brewing water can have a direct effect on beer flavor. They can influence the reactions that take place during the various stages of brewing, or they can interact with other flavor components. (See Table 4.3.)

Carbonate

In addition to H_3O^+ and OH^-, a number of other ions are generally present in water. Of these, the more common are the carbonate ion CO_3^{2-} and its close relative, the bicarbonate ion (also called hydrogen carbonate ion) HCO_3^-. Ions of the carbonate family can be thought of as originating with carbon dioxide (CO_2).

$$CO_2 \text{ (gas)} + H_2O \text{ (liquid)} \rightarrow H_2CO_3 \text{ (dissolved carbonic acid)}$$

$$H_2CO_3 + H_2O \rightarrow H_3O^+ + HCO_3^- \text{ (dissolved bicarbonate ion)}$$

$$HCO_3^- + H_2O \rightarrow H_3O^+ + CO_3^{2-} \text{ (dissolved carbonate ion)}$$

The fraction of these ions present as CO_3^{2-}, HCO_3^-, or H_2CO_3 depends on the pH of the water. As the pH gets higher (more basic) the CO_3^{2-} fraction increases and the HCO_3^- and H_2CO_3 fractions decrease. This is an application of Le Châtelier's principle; higher pH means less H_3O^+ so the reaction moves to make more H_3O^+, which results in more CO_3^{2-}. A one unit increase in the pH of the water changes the ratio of carbonate to bicarbonate molar concentrations by a factor of 10. The concentrations of carbonate and bicarbonate ions are equal at a pH of 10.3. At a pH of 11.3 the concentration of carbonate is ten times as high as that of bicarbonate. At a pH of 9.3 the concentration of carbonate is ten times lower than that of bicarbonate. At all stages of brewing, the pH is such that bicarbonate ion is present in much greater concentration than that of carbonate ion. At the pH of beer, there is about one

> **MEASURING ALKALINITY**
>
> Alkalinity is measured by the amount of hydrochloric acid (HCl) that can be added to the water before it reaches a pH of (usually) 4.2. The number of millimoles (1000th of a mole) of HCl per liter of water is the alkalinity in milliequivalents per liter. Sometimes this result is multiplied by 50 mg/mmol (mg of $CaCO_3$ neutralized by one mmol of HCl) to give the alkalinity in milligrams of $CaCO_3$ per liter (ppm $CaCO_3$). Reporting alkalinity as though it were due entirely to $CaCO_3$ is conventional but misleading; most alkalinity is due to the bicarbonate ion.

million times as much bicarbonate as carbonate. Pale lager beer owes some of its character to water with a low bicarbonate content. Higher concentrations of bicarbonate in brewing liquor are considered suitable for darker or more full-flavored styles. The amount of acid that can be neutralized by the carbonate, bicarbonate, and certain other ions in water is called the alkalinity of the water. Water that is excessively alkaline leads to high mashing and fermentation pH, which can affect the quality of the beer.

Hard and Soft Water

Water that is high in ions with a +2 or +3 charge is called hard water. The opposite is **soft water**. In brewing, the relevant ions are calcium ions (Ca^{2+}) and magnesium ions (Mg^{2+}). Calcium carbonate ($CaCO_3$) is insoluble in water, so if calcium ions and carbonate ions are present in appreciable concentration, they will find each other, form a solid, and drop out of solution (**precipitate**, in chemistry jargon). Magnesium carbonate is much more soluble, so it has less of a tendency to precipitate from water. Hard water can be desirable for some styles of beer, like British ales; soft water is preferred for others, like Pilsner lager. Hardness can be a nuisance in equipment that heats or boils water. Heating can cause calcium carbonate to precipitate, making a stony deposit called **pipe scale** on the hot surfaces. Pipe scale restricts the flow of heat in heating and cooling equipment and can even block pipes.

> **MEASURING HARDNESS**
>
> **Hardness** in water is caused by dissolved ions with a charge of +2 or more. The major contributors to hardness in drinkable water are magnesium ion (Mg^{2+}) and calcium ion (Ca^{2+}). The classic way to measure hardness is to treat a water sample with a reagent called ethylenediamine tetraacetate (EDTA, Fig. 4.6), which reacts with these ions. The number
>
> *(continued)*

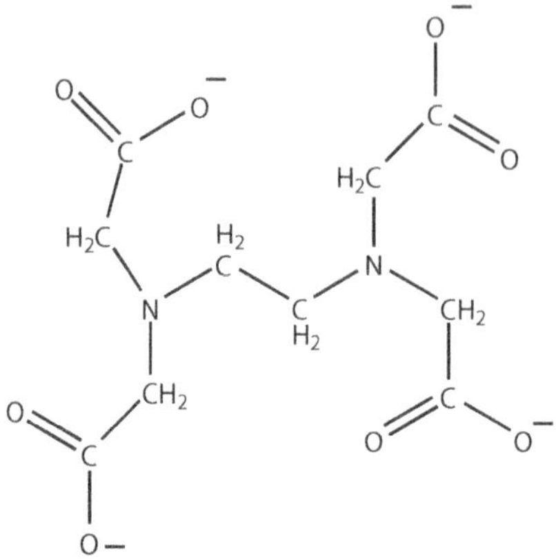

Figure 4.6 EDTA.

of moles of EDTA needed to react with all the hardness ions is equal to the number of moles of ions. The molar concentration of the ions, which is the number of moles of ions divided by the water volume in liters, is a measure of the hardness of the water. Usually the molar concentration is multiplied by 2000 to give a concentration in milliequivalents per liter (mEq/L), where a milliequivalent is 0.001 mole of charge (one mole of Ca^{2+} or Mg^{2+} contributes two thousand milliequivalents).

Hardness is often reported in milligrams per liter (mg/L—also called ppm) or grains per gallon (GPG, 1 grain/gal = 17.12 mg/L), often with the units omitted. In both mg/L and GPG the hardness is calculated as though it were all due to calcium ions. The mass of calcium carbonate ($CaCO_3$) that would contain this much calcium ion is calculated and serves as the mass (milligrams or grains).

Note that converting from total hardness to concentration of a single ion assumes that all the hardness comes from that ion. To convert from the units on the left to those on top, multiply by the table entry.

Hardness Unit Conversions

From	ppm Mg^{2+}	ppm Ca^{2+}	ppm Hardness as $CaCO_3$	mEq/L	Hardness gr/gal $CaCO_3$
ppm Mg^{2+}	1	—	4.118	0.0823	0.2406
ppm Ca^{2+}	—	1	2.497	0.0499	0.1459
ppm Hardness $CaCO_3$	0.2428	0.4004	1	0.0200	0.0584
mEq/L	12.152	20.039	50.043	1	4.1180
Hardness gr/gal $CaCO_3$	4.147	6.855	17.12	0.2428	1

> **RESIDUAL ALKALINITY**
>
> The hardness ions Ca^{2+} and Mg^{2+} can lower the pH of wort, counteracting to some extent the effect of alkalinity. This is accounted for in the residual alkalinity (RA):
>
> $$RA = TA - \frac{[Ca^{2+}]}{3.5} - \frac{[Mg^{2+}]}{7}$$
>
> where TA is the total alkalinity, and $[Ca^{2+}]$ and $[Mg^{2+}]$ are the concentrations of these ions, all in milliequivalents per liter. Water with excessive residual alkalinity must be acidified to avoid a high mashing pH.

Ions and Mash pH

Finished beer has a pH of around 4. The more critical issue is the pH during mashing, which affects the functioning of the mashing enzymes. Two ions in brewing water can exert a major influence on the mash pH. These are the bicarbonate ion (HCO_3^-), which raises the pH, and the calcium ion (Ca^{2+}) and, to a smaller extent, the magnesium ion (Mg^{2+}), which lower the pH.

The bicarbonate ion reacts with hydronium ion as shown below, removing it from solution and lowering its concentration. Lower H_3O^+ means less acidity and higher pH.

$$HCO_3^- + H_3O^+ \rightarrow CO_2 + 2H_2O$$

Calcium ion influences mash pH by a less direct reaction. Many compounds in living and dead cells (like grain seeds) are attached to the hydrogen phosphate ion, HPO_4^{2-}. The hydrogen phosphate ion is essentially neutral on its own, but it can form an insoluble compound with calcium ion, and a less insoluble compound with magnesium ion:

$$3Ca^{2+} + 2HPO_4^{2-} \rightarrow Ca_3(PO_4)_2(solid) + 2H^+$$

The H^+ ions react immediately with water to give hydronium ions (H_3O^+), decreasing the pH.

4.5 WATER TREATMENT

Tap, well, or river water usually needs to be treated to make good beer. Most available water is harder and more alkaline than the brewer wants. Disinfectants like chlorine (Cl_2) or chloramine (NH_2Cl) need to be removed. If the

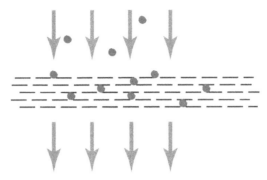

Figure 4.7 Filter. (See color insert.)

water contains excessive amounts of iron(II) ion (ferrous ion: Fe^{2+}), it will need to be removed. **Purification** involves separation of water from the stuff we want to remove. Any type of purification requires energy to overcome the spontaneous tendency for substances to mix rather than to separate.

Filtration

Water, when it leaves the water company, is quite clean and clear. As it flows through miles of pipes, the water is likely to pick up bits of rust, pipe scale, and other particles. Particles are removed by filters that operate on the principle that the particles are hundreds of thousands of times larger than the molecules of water. A filter has pores or channels that allow the water to go through but are too small for the particles (Fig. 4.7). The filtering element can be a bed of clean sand or other mineral, porous paper (like a coffee filter), or porous plastic. After the filter picks up its capacity of particles, it has to be cleaned or replaced. Filters only remove solid particles from the water. They do not remove individual dissolved ions or other dissolved materials, like chlorine, that are mixed with the water at the molecular level.

Iron Removal

One way to remove Fe^{2+} (iron(II), ferrous ion) from water is to pull an electron from the Fe^{2+}, which converts it to Fe^{3+} (iron(III), ferric ion). Pulling electrons from something is called **oxidation**, because the usual way to do it is to treat with oxygen. Oxygen atoms grab electrons to fill their valence shells, producing oxide ions (O^{2-}):

$$4Fe^{2+} + O_2 \rightarrow 4Fe^{3+} + 2O^{2-}$$

Fe^{3+} has an extremely insoluble hydroxide, $Fe(OH)_3$, which is the all-too-familiar reddish compound that we know as a component of rust. The Fe^{3+} ions find the traces of hydroxide ion (OH^-) always present in water, forming a solid that is allowed to settle out. If the iron(II) ion is not removed from the water, this reaction is likely to occur in some inconvenient place, leaving a slimy coating on the equipment.

Reverse Osmosis

There are membranes with pores that admit only water molecules, which are quite small, but do not admit ions, large molecules, or microscopic organisms. Water will flow spontaneously from the clean side to the dirty side, which is not what we want. This spontaneous flow is called **osmosis**. To purify water, we have to get clean water to flow from the dirty side to the clean side, making more clean water. This nonspontaneous flow is called **reverse osmosis**. In order to get the water to undergo reverse osmosis it must be driven by pressure. The higher the concentration of dissolved materials is on the dirty side, the higher is the pressure needed to drive clean water through the membrane. Practical reverse osmosis units use membranes configured as thin tubes to give a large surface for the water to cross.

OSMOSIS: A CLOSER LOOK

Suppose we get a thin film whose pores are so small that they will pass only certain molecules and not others. Such a film is not the same as a filter; it is a **semipermeable membrane**. In Figure 4.8 the membrane divides a container into two compartments. The right compartment has only molecules that can go through the membrane; the left compartment

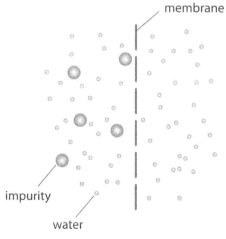

Figure 4.8 Osmosis. (See color insert.)

(*continued*)

also includes some impurity molecules that will not go through the membrane. Whenever any molecule in the right compartment hits a pore, it goes through. But some of the molecules on the left do not go through. As a result, there is a greater flow of molecules from the right than from the left. That is, liquid from the clean side tends to move into the impurity side.

This flow, called **osmosis**, is exactly the opposite of what we want. We want to make dirty water clean; we do not want our clean water to go to the dirty side. If we apply pressure to the water in the left compartment, we increase the flow from left to right. At a certain pressure, called the **osmotic pressure**, the flow from the left becomes equal to the flow from the right. At this pressure, there is no net flow. The osmotic pressure depends on the concentration of particles that will not pass through the membrane. There is still flow in both directions, but because it is the same in both directions, we don't see any change. If the pressure applied on the left is greater than the osmotic pressure, flow goes from left to right; that is, we drive pure liquid through the membrane leaving impurities behind. This is called **reverse osmosis**. The water that remains in the compartment on the left has a higher concentration of impurities than before, so the osmotic pressure increases. During freshwater purification about 75% of the inlet water is driven through the membrane before the pressure becomes too high to make it worthwhile. The remaining impure water can be used for procedures that don't need high quality water.

Living cells are essentially little bags of impure water surrounded by a semipermeable membrane. If the water on the outside has a low concentration of impurities, water will flow into the cell. If the water on the outside has a high concentration of impurities, the water will flow out of the cell. These conditions place *osmotic stress* on the cell to which it must respond if it is to survive. Osmotic stress can make fermentation of very concentrated sugar solutions difficult.

Ion Exchange

Ion exchange involves passing the water through a bed of solid granules that hold on to ions we want to get rid of and release ions that we don't mind. Cation (CAT eye on) (positive ion) exchangers grab positive ions like Ca^{2+}, Mg^{2+}, Fe^{2+}, and Mn^{2+} and release either two H^+ or two Na^+ ions to maintain the same charge. Anion (ANN eye on) exchangers grab negative ions like SO_4^{2-} and release enough negative ions, two OH^- or two Cl^- ions, to maintain the charge. Systems that release H^+ and OH^- ions are called **demineralizers** or **deionizers**, because these ions combine to form pure water (Fig. 4.9). Systems

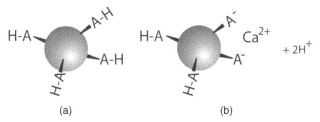

Figure 4.9 (a) Cation exchange granule and (b) cation exchange granule with calcium ion.

that release Na^+ and Cl^- ions are called **water softeners**. Demineralizers are regenerated by treatment of the cation exchanger with a concentrated acid solution and by treatment of the anion exchanger with concentrated base solution. Softeners are easier to regenerate; both parts are treated with the same concentrated salt (NaCl) solution. Because of this, the anion and cation exchange resins in a softener can be mixed together in a single vessel. Demineralization is most applicable to brewing liquor. Softening is cheaper and is satisfactory for cooling, heating, or wash water.

Activated Carbon

Coal or materials derived from plant or animal parts, like coconut shells or olive pits, are heated without air to give a form of carbon called **coke** if coal is used, or **charcoal** if plant or animal parts are used. These are then heated with a substance that reacts with carbon to make many little holes called pores, a process called activation. Steam, carbon dioxide, or oxygen can be used for activation. The resulting **activated carbon** has a complex network of pores whose walls grab uncharged molecules by dispersion forces or by chemical reaction with the carbon. This is the principle of water purifiers like the Brita®. Chlorine (Cl_2), chloramine (NH_2Cl), and carbon-based materials can be removed. Activated carbon is often used to clear water of substances that can give a bad flavor to the beer.

Oxygen Removal

Some breweries try to minimize contact with oxygen from the air at all stages of the brewing process. Dissolved oxygen can be removed from water by letting the water trickle down a bed of beads while nitrogen or carbon dioxide flows up through the bed. The solubility of a gas in a liquid depends on the concentration of that gas in contact with the liquid. The stripping gas flowing up the bed has very little oxygen, so the oxygen tends to come out of solution and go into the gas stream. This is another example of Le Châtelier's principle.

Boiling

When alkaline water is heated, especially if it is boiled, bicarbonate ions (HCO_3^-) undergo the reaction $2HCO_3^- \rightarrow CO_3^{2-} + CO_2 + H_2O$. This can increase the carbonate ion concentration to the point that $CaCO_3$ drops out of solution, making the water softer and less alkaline. Hardness that is removed by boiling is called **temporary hardness**. Hardness that remains after boiling is called **permanent hardness**.

In the past, brewing water was often boiled, then chilled to remove excess hardness and alkalinity. The solid carbonates were allowed to settle out. Boiling is seldom used today to soften water in commercial breweries because of the high cost of the energy to heat the water. Some homebrewers boil their water to soften and sanitize it.

Water Adjustment

Some brewers add ionic compounds (salts) to brewing water to make the water more suitable for the style of beer they are brewing. All such additions should be of food grade quality. Some commonly used additives include calcium chloride dihydrate ($CaCl_2 \cdot 2H_2O$), calcium sulfate dihydrate (gypsum, $CaSO_4 \cdot 2H_2O$), calcium carbonate (limestone, also called chalk, $CaCO_3$) and magnesium sulfate heptahydrate (Epsom salt, $MgSO_4 \cdot 7H_2O$). The compounds with dot-H_2O in the formula are **hydrates**. A hydrate is a compound with one or more molecules of water bound within the structure. The coefficient of water indicates the number of water molecules incorporated into the formula. Some of these compounds are also available in anhydrous form, that is, without the water. When the compound is weighed, the water goes on the scale with the rest of it, so the mass of the water must be taken into account in using these compounds. (See Table 4.4.)

TABLE 4.4 Ion Content of Salts

Name	Formula	Ions	
Anhydrous calcium chloride	$CaCl_2$	Ca^{2+}: 36.1%	Cl^-: 63.9%
Calcium carbonate	$CaCO_3$	Ca^{2+}: 40.0%	CO_3^{2-}: 60.0%
Calcium chloride dihydrate	$CaCl_2 \cdot 2H_2O$	Ca^{2+}: 27.3%	Cl^-: 48.2%
Calcium sulfate dihydrate	$CaSO_4 \cdot 2H_2O$	Ca^{2+}: 23.3%	SO_4^{2-}: 55.8%
Magnesium sulfate heptahydrate	$MgSO_4 \cdot 7H_2O$	Mg^{2+}: 9.9%	SO_4^{2-}: 39.0%
Sodium bicarbonate	$NaHCO_3$	Na^+: 27.4%	HCO_3^-: 72.6%
Sodium chloride	$NaCl$	Na^+: 39.3%	Cl^-: 60.7%

HYDRATES: A CLOSER LOOK

The mass fraction of a particular ion in a compound, such as those used for water adjustment, is the mass of the ion in a mole of compound divided by the molar mass of the compound, including water. To determine the mass fraction of magnesium in Epsom salt, we take the mass of the magnesium in one mole (24.31 grams) and divide by the molar mass of $MgSO_4 \cdot 7H_2O$, calculated below:

Mg	1×24.10 g/mol	24.31 g/mol
S	1×32.06	32.06
O	4×16.00	64.00
water	7×18.015	126.11
		246.47 g/mol

The mass fraction is 24.3 g/mol divided by 246.47 g/mol, which comes to 0.099 or 9.9%. Suppose we want to add enough Epsom salt to one gallon of water to give water with 62 mg/L magnesium ion. A gallon is 3.785 liters, so we need 3.785 L × 62 mg/L = 235 mg = 0.235 gram. To get the mass of Epsom salt containing 0.235 g of magnesium, we divide the mass of magnesium by the mass fraction: 0.235 g ÷ 0.099 = 2.37 g.

BIBLIOGRAPHY

Briggs, Dennis E.; Boulton, Chris A.; Brookes, Peter A.; and Stevens, Roger. *Brewing Science and Practice*. CRC Press, 2004, Chap. 3. All about water, how it is used, tested, and disposed of.

Eumann, M. Water in Brewing. In *Brewing: New Technologies*, Bamforth, C. W., Ed. CRC Press, 2006, pp. 183–207. Standards and treatment methods for brewery water for various purposes.

Kohlbach, P. *Monatschrift für Brauwissenschaft*, **1953**, 6(5): 45. A hard-to-get article in German documenting the residual alkalinity concept. Uses strange European units. The reference by Palmer is a more practical source of information.

Palmer, John J. *How to Brew*. Brewers Publications, 2006, Chap. 15. A readable discussion of water issues.

Taylor, David G. Water. In *Handbook of Brewing*, 2nd Ed., Priest, Fergus G. and Stewart, Graham G. Eds. Taylor and Francis, 2006, Chap. 4. How individual ions affect beer.

QUESTIONS

4.1. What is the chemical formula for water?

4.2. If a sample of water has 1000 hydrogen atoms, how many molecules of water does the sample have?

4.3. Identify the intermolecular forces that attract molecules of water to one another.

4.4. Identify three trace substances in water that are needed for the brewing process.

4.5. Identify the following pH values as acidic, neutral, or basic: 4, 6, 7, 9.

4.6. Is beer acidic, basic, or neutral?

4.7. Does the addition of an acid to water make the pH increase, decrease, or stay the same?

4.8. Complete the acid–base equation: $HClO_4 + H_2O \rightarrow$

4.9. What is hard water?

4.10. Name two acids and two bases.

4.11. Identify two problems with water that make water treatment necessary.

4.12. What is activated carbon? For what is it used?

4.13. Describe filtration, reverse osmosis, and ion exchange.

4.14. Explain the difference between softening and demineralization.

4.15. A material whose isoelectric point is 5.8 is added to a solution whose pH is 4.2. What can we say about the electric charge on particles of the material?

*4.16. Calcium carbonate is sometimes used to raise the mashing pH and calcium chloride to lower it. Explain how these compounds affect the pH. Show chemical equations.

*4.17. Calculate the pH of a solution that is 3×10^{-4} mol/L in H_3O^+.

*4.18. Calculate the total hardness in ppm $CaCO_3$ equivalent for water that has 25 ppm magnesium ion and 125 ppm calcium ion.

***4.19.** If the water in Question 4.18 has a total alkalinity of 185 ppm $CaCO_3$, calculate the residual alkalinity in mEq/L.

****4.20.** A 50 mL sample of water is analyzed with EDTA: 0.0375 mmol (3.75×10^{-5} mole) of EDTA is required. Calculate the total hardness in ppm $CaCO_3$ equivalent.

Outside view showing silo and railroad car. Yuengling Beer Company, Pottsville, Pennsylvania.

CHAPTER 5

INTRODUCTION TO ORGANIC CHEMISTRY

Carbon is easily the most versatile of all elements. There is no evident limit to the number of carbon atoms that can be connected together in chains, branched chains, and rings to form molecules. This variety is enhanced by the inclusion of other nonmetals, especially oxygen and nitrogen in the molecules. Compounds with carbon–hydrogen or carbon–carbon bonds are called **organic** compounds; their chemistry is called **organic chemistry**. Organic compounds form the basis for all life on earth. Much of the chemistry of beer is organic chemistry.

5.1 STRUCTURAL FORMULAS

All but the simplest organic compounds share their formulas with other compounds. Different compounds with the same chemical formula are called **isomers**. The compounds shown in Figure 5.1 all have the formula C_4H_8O. Isomers may have completely different properties from one another. To be specific about which isomer we are dealing with, we will usually represent organic compounds with **structural formulas**. A structural formula is a representation that shows how the atoms in a molecule are connected. There are many versions of structural formulas; the variety can be confusing. A **Lewis structure** is a drawing that shows all the atoms represented by their symbols. Covalent bonds are represented by lines, and unshared electrons may be included as dots. Sometimes the bond angles are approximated in the drawing,

---

The Chemistry of Beer: The Science in the Suds, First Edition. Roger Barth.
© 2013 John Wiley & Sons, Inc. Published 2013 by John Wiley & Sons, Inc.

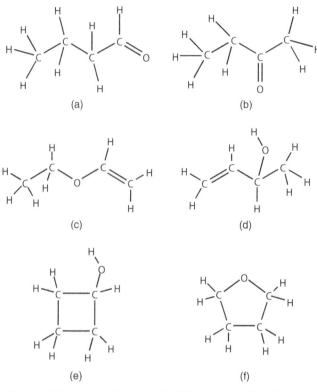

Figure 5.1 Structural formulas: (a) butanal, (b) butanone, (c) ethoxyethene, (d) but-3-ene-2-ol, (e) cyclobutanol, and (f) tetrahydrofuran.

and sometimes the bonds are drawn as right angles, irrespective of their actual angles. Sometimes the structure incorporates a three-dimensional effect by giving the bonds a perspective appearance.

Lewis structures can become cumbersome and confusing, so a number of simplifications are used. These are based on the facts that carbon always takes four bonds and that hydrogen only takes one bond. So we can write a condensed structure for ethanol as CH_3–CH_2–O–H, or just CH_3CH_2OH. It is understood that the hydrogen atoms written after each carbon are attached to that carbon only and that each carbon atom is bonded to the next nonhydrogen atom. In this way we can easily distinguish between ethanol and its isomer, dimethyl ether CH_3–O–CH_3. Condensed structures are a convenient way to fit a structure into a line of text, but they are not well adapted to showing features like branches, rings, or any nonhydrogen atoms that are not in a straight line. For example, acetone (Fig. 5.2) would have to be written as CH_3–C(=O)–CH_3, where the parentheses are an awkward device to show that the oxygen and the following carbon atom are both bound to the middle carbon atom. In the semicondensed structure (Fig. 5.3b) we show all atoms but we omit many of the lines representing bonds to hydrogen atoms. All other

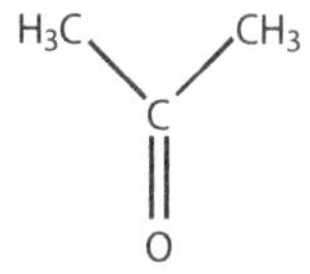

FUNCTIONAL GROUPS   91

Figure 5.2   Acetone.

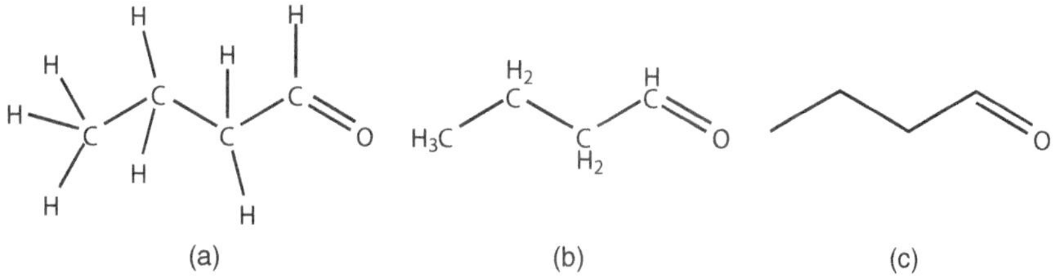

Figure 5.3   Butanal: (a) Lewis structure, (b) semicondensed structure, and (c) skeletal structure.

bonds are shown. We write the hydrogen atoms that are bound to a particular carbon near that carbon, but without the symbols for bonds.

The **skeletal structure** (Fig. 5.3c) is a simplification of the Lewis structure in which the symbols for the carbon atoms are omitted. Hydrogen atoms (along with their bonds) attached to carbon atoms are omitted. Usually bond angles are represented approximately accurately. The end of a bond that is not marked as some other atom is understood to be a carbon atom. Each carbon atom must have four bonds, so any bonds not explicitly shown are understood to be connected to hydrogen atoms. This convention can give a clear picture of the shape of the molecule and tends to highlight the interesting parts, that is, the groups with nitrogen or oxygen atoms. Figure 5.3 compares Lewis, semicondensed, and skeletal structures for the same compound. Many chemists are not comfortable with the angular zig-zags of skeletal structures, so they may prefer to use strict skeletal notation largely for rings. We will adopt this approach and we will show both carbon and hydrogen atoms outside of rings. We will often put hydrogen atoms near the carbon atoms to which they are bound without showing the bonds (Fig. 5.4). Hydrogen atoms attached to carbons in rings will be omitted except to show certain three-dimensional features. Understanding the basics of chemical notation is an important part of chemical literacy.

5.2   FUNCTIONAL GROUPS

The number of ways to assemble atoms into organic molecules is nearly infinite. It is convenient to think about the molecules on the basis of collections of atoms that form parts of the molecules. These collections are called **groups** or **functional groups**. Families of compounds are classified by the functional

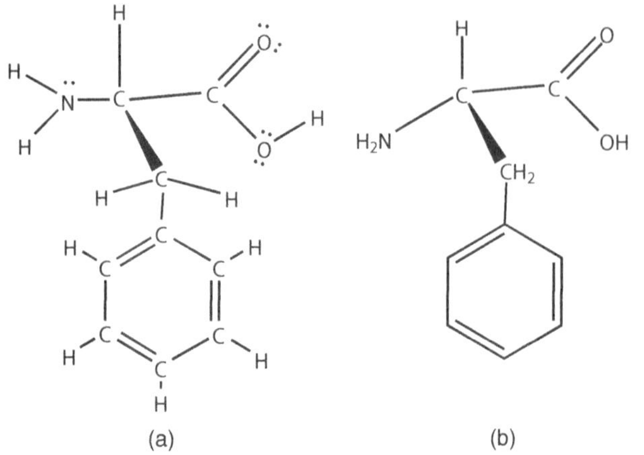

Figure 5.4 Phenylalanine: (a) Lewis structure, and (b) modified skeleton.

groups they contain. For example, a molecule with an –OH group attached to a carbon atom that has no other oxygen attached is an **alcohol**. Each group gives particular behavior to the molecule or to part of the molecule. This approach is useful, but not perfectly exact. Molecules with the same functional groups can still differ in many ways, and sometimes molecules display unexpected behavior not predicted on the basis of the functional groups. Many biological molecules have multiple functional groups and belong to several families at once. When multiple functional groups are present, the compound is named after the highest priority group. Lower priority groups are called **substituents**. The order of priorities for common groups is carboxylic acid > ester > aldehyde/ketone > alcohol > amine > alkene > alkyne > alkane > ether.

Alkanes and Alkyl Groups

An **alkane** is a compound containing only carbon and hydrogen atoms and only single bonds. A portion of a molecule that has only carbon and hydrogen and only single bonds is called an **alkyl group**. Chemists tend to think of alkanes as the absence of functionality. Because C–H and C–C bonds have low to no polarity, alkanes are essentially nonpolar. Regions of a molecule that have only alkyl groups are **hydrophobic**; that is, they avoid polar molecules like water, and they interact best with nonpolar molecules or with nonpolar regions of molecules. Some examples of alkyl groups are shown in Figure 5.5 in different representations. Each alkyl group is shown with an extra bond that serves as the attachment point to the rest of the molecule. The difference between the *n*-propyl (also called 1-propyl) group and the isopropyl (also called 2-propyl) group is the point of attachment. The *n*-propyl group attaches to the rest of the molecule by its end carbon and the isopropyl group attaches by its middle carbon. The C–C single bonds in alkyl groups

FUNCTIONAL GROUPS   93

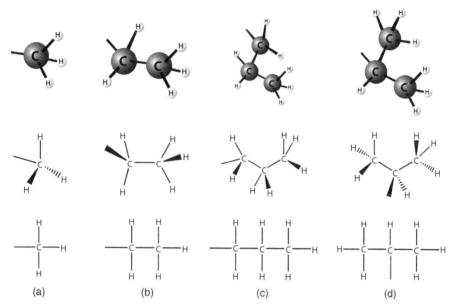

Figure 5.5 Alkyl groups: (a) methyl group, (b) ethyl group, (c) *n*-propyl group, and (d) isopropyl group.

can rotate freely, so the shapes of molecules with alkyl groups are not usually fixed. As chains of carbon atoms become longer, they tend to adopt a zigzag shape. Alkyl groups are present in most organic compounds that are important for beer and brewing. Most families of organic compounds have a characteristic functional group attached either to an alkyl group or to a hydrogen atom. Sometimes when we want to refer to any general alkyl group or hydrogen atom, we use the symbol R–. So the general formula for an alkane can be written R–H.

NAMING ALKANES

The root name for an alkane consists of a prefix representing the code for the number of carbon atoms in the longest continuous chain, followed by the suffix –ane. Rings get a cyclo prefix before the number prefix. If there are shorter chains sticking out of the main chain, a prefix is added consisting of a number telling where the short chain attaches followed by the short chain length code, followed by -yl. If there is more than one of a particular chain length, a prefix di-, tri-, tetra-, and so on is applied.

(continued)

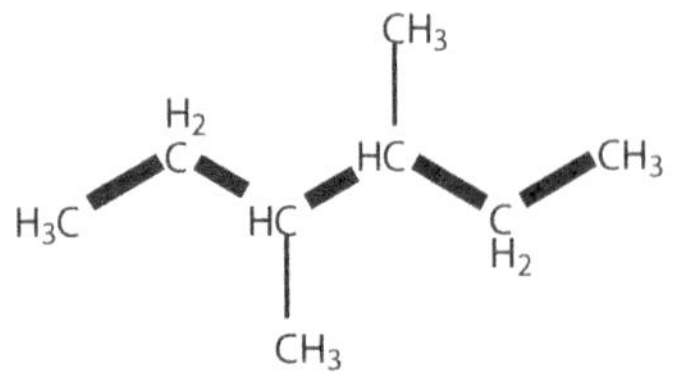

Figure 5.6 3,4-Dimethyhexane.

Carbons	1	2	3	4	5	6	7
Code	meth-	eth-	prop-	but-	pent-	hex-	hept-

Carbons	8	9	10	11	12	16	18
Code	oct-	non-	dec-	undec-	dodec-	hexadec-	octadec-

The molecule in Figure 5.6 has a six-carbon main chain, shown in bold. So the root is hexane. There are two one-carbon side chains attached to the third and fourth carbon atom of the main chain. The full name is 3,4-dimethylhexane.

Alkenes/Aromatic Compounds/Alkynes

The carbon–carbon double bond, in which the two carbon atoms share two pairs of electrons, is considered a functional group. A compound with one or more C=C bonds is either an **alkene** or an **aromatic** hydrocarbon. **Terpenes**, a class of alkenes, (Fig. 5.7) provide hop aromas. The distinction between an alkene and an aromatic hydrocarbon is that aromatic compounds have rings in which the atoms are connected by an alternating pattern of single and double bonds as shown in Figure 5.8.

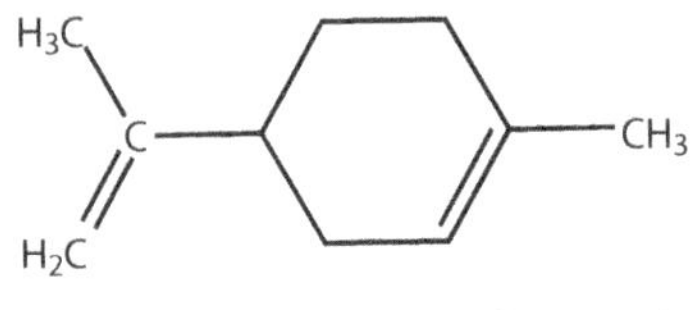

Figure 5.7 Limonene (terpene).

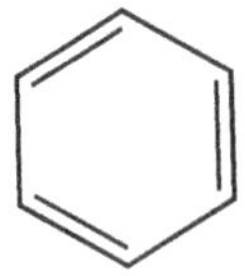

Figure 5.8 Benzene (aromatic).

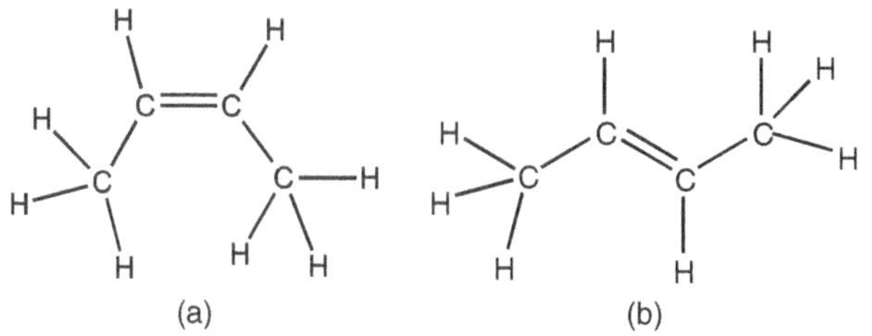

FUNCTIONAL GROUPS   95

Figure 5.9 Geometric isomers: (a) *cis*-2-butene and (b) *trans*-2-butene.

Figure 5.10 Addition reaction.

There is no rotation about a double bond, and all atoms attached to the double bond lie in a flat plane. Because of this, some molecules can differ by which side of the double bond the attached groups are on. In the compounds in Figure 5.9, the only difference is that *cis*-2-butene on the left has both hydrogen atoms attached to the C=C carbons on the same side (up) and *trans*-2-butene on the right has one hydrogen up and one down. Such compounds are called **geometric isomers** or cis–trans isomers. Geometric isomers are distinguished by the arrangement of groups on the same side or across from one another with reference to some feature of the molecule. Geometric isomers are different compounds with different physical, chemical, and biological properties, which is chemistry jargon that means they may look different, have different melting and boiling points, react in different ways, and taste and smell different.

The characteristic reaction of a double bond (between any atoms) is the **addition** reaction. Figure 5.10 shows addition of X–Y across the double bond between E and D. One of the shared pairs of electrons between E and D separates. Each of these two electrons forms a shared pair with one of the electrons forming the X–Y bond. Because they are subject to the addition reaction, alkenes are more reactive than alkanes; most of them react slowly with oxygen in air. The reason for the distinction between alkenes and aromatic compounds is that the addition reaction is much less likely in aromatic compounds. As a result, aromatic compounds are often more stable than alkenes. Aromatic rings with nitrogen and oxygen atoms serve as identifying tags providing information to living cells. The bases, like adenine (Fig. 5.11), that carry genetic information in DNA are aromatic compounds that function in this way.

Compounds with a C≡C triple bond are called alkynes (al-KINE). Alkynes are not usually important in beer.

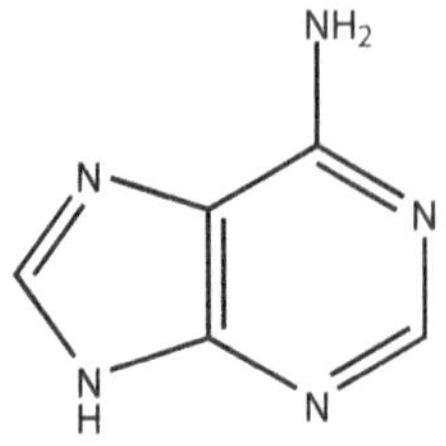

Figure 5.11 Adenine.

NAMING ALKENES AND ALKYNES

The root name of an alkene consists of the code for the number of carbon atoms in the longest chain containing the double bond, plus the suffix -ene. There is a prefix with the carbon number preceding the double bond. If there are side chains or anything else with numbers, the number goes right before the -ene instead of as a prefix.

In Figure 5.12 the longest chain with the double bond has five carbons, so the chain length code is pent-. The double bond is after carbon number 2 (we count from the end nearest the functional group) so the -ene suffix is marked with the number 2. There is a one-carbon side chain (a methyl group) on carbon 3: 3-methylpent-2-ene.

If there is a C=C double bond on a substituent, it gets the prefix for the substituent chain length with an -enyl suffix.

Alkynes are named in the same way, but the suffix is -yne instead of -ene.

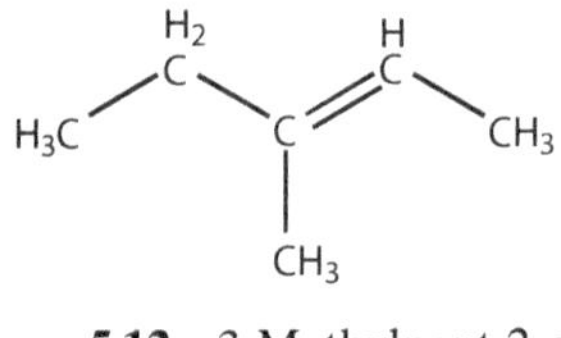

Figure 5.12 3-Methylpent-2-ene.

Alcohols

An **alcohol** is a compound with an –OH group attached to a carbon that does not have a bond to any other O or N atoms. (Fig. 5.13) The general formula for an alcohol is R–OH. Comparing this to the formula for water, H–OH, an alcohol can be thought of as water in which an alkyl group was substituted for one of the hydrogen atoms. When R represents the ethyl group, the formula is CH_3CH_2OH, ethanol or ethyl alcohol, the compound that makes alcoholic beverages what they are. Molecules with two –OH groups on the same carbon atom are usually unstable. The –OH group is highly polar and can engage in hydrogen bonding with water or other alcohol molecules. The presence of an

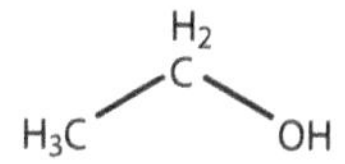

Figure 5.13 Ethanol.

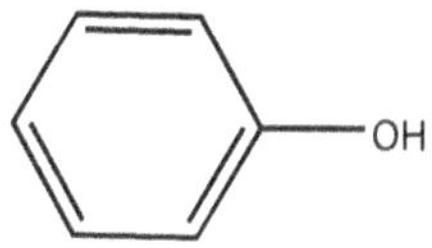

Figure 5.14 Phenol.

–OH group makes a molecule or a region of the molecule **hydrophilic**. Hydrophilic molecules or regions tend to interact with water or other polar molecules and to avoid hydrophobic (or nonpolar) molecules or regions. Alcohols with one to three carbons are completely soluble in water. Most alcohols are much less likely than water to donate or accept hydrogen ions, so acid–base behavior is less important. An alcohol in which the –OH group is directly attached to an aromatic ring (Fig. 5.14) is called a **phenol** (FEE nole or feh NOLE) (named after a compound called phenol). Phenols are about one million times more acidic than other alcohols. Alcohols can react with one another and with other compounds with –OH groups (like carboxylic acids— more later) in a **dehydration** reaction: R–OH + HO–R → R–O–R + H_2O, where R–OH and HO–R are alcohols and R–O–R is an **ether**. Dehydration is an important way for molecules to attach to one another to form chains. R–O–R is the general formula for an ether. The reverse of this reaction, called **hydrolysis**, is what makes starches into fermentable sugars.

NAMING ALCOHOLS

Alcohols are named by adding the suffix -anol to the code for the number of carbon atoms in the longest chain containing the carbon with the –OH group. A numeral prefix indicates the position of the –OH group. If the –OH is a substituent (side group) because of higher priority functional groups, it is represented by the prefix hydroxy-.

In Figure 5.15 the longest chain has four carbons so the prefix is but- and the suffix is -anol. The –OH is on carbon 2 (count from the near end), so the name is 2-butanol.

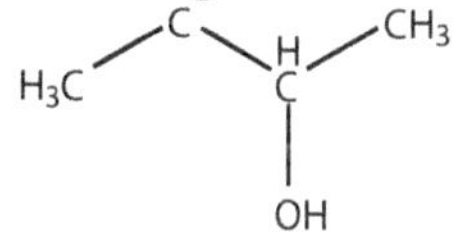

Figure 5.15 2-Butanol.

98 INTRODUCTION TO ORGANIC CHEMISTRY

Figure 5.16 Tetrahydrofuran.

Ethers

An **ether** has two alkyl groups connected to an oxygen atom. The general formula is R–O–R, which is called the ether group. The ether molecule is bent at the oxygen atom like the water and alcohol molecules, so it has an oxygen end and an alkyl group end. Ethers are polar. Cyclic ethers have the oxygen incorporated into a ring of carbon atoms (Fig. 5.16). Because ethers have no O–H bond, they do not form hydrogen bonds with one another. Most ethers are not very soluble in water. The exceptions are some cyclic ethers. Ethers are not very reactive under moderate conditions, but some of them are highly flammable. Some ethers react slowly with air to make **peroxides** (ROOR), many of which are explosive.

NAMING ETHERS

Ethers are commonly named by naming the two groups attached to the oxygen followed by "ether." Some examples would be methylpropyl ether, or diethyl ether. The official way to name an ether is to consider the smaller or lower priority –OR group as a side group. The side group is named by the code for the number of carbon atoms with a suffix -oxy.

In Figure 5.17 there is a methyl group and an ethyl group attached to the oxygen, so the common name is methylethyl ether. The official name is methoxyethane.

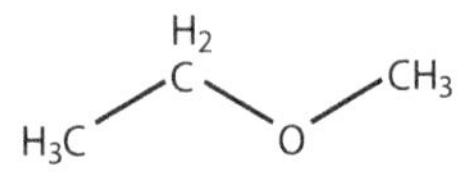

Figure 5.17 Methoxyethane.

Amines

An **amine** (um EEN) is a compound with the **amino** group. The amino group is $-NH_2$, although sometimes one or both of the hydrogen atoms on the nitrogen are substituted with something else, such as an alkyl group. If neither of the two H atoms is substituted, the amine is a primary amine. If one of the H atoms is substituted, the amine is a secondary amine, if two are substituted, it

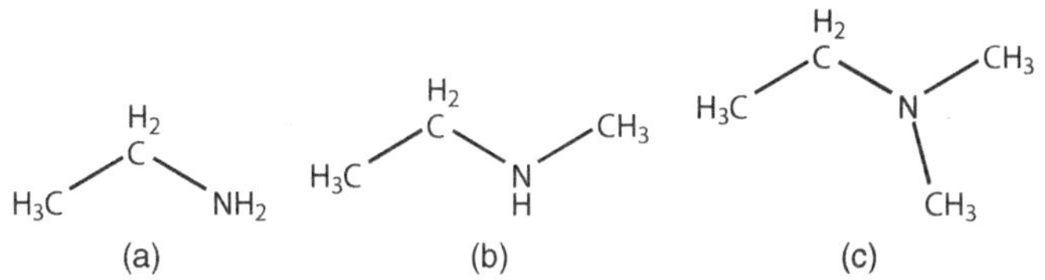

Figure 5.18 Amines: (a) ethylamine, primary amine; (b) *N*-methyl ethylamine, secondary amine; and (c) *N,N*-dimethyl ethylamine, tertiary amine.

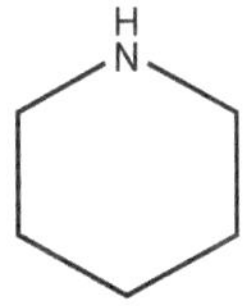

Figure 5.19 Piperadine secondary amine.

Figure 5.20 Ethylamine as base.

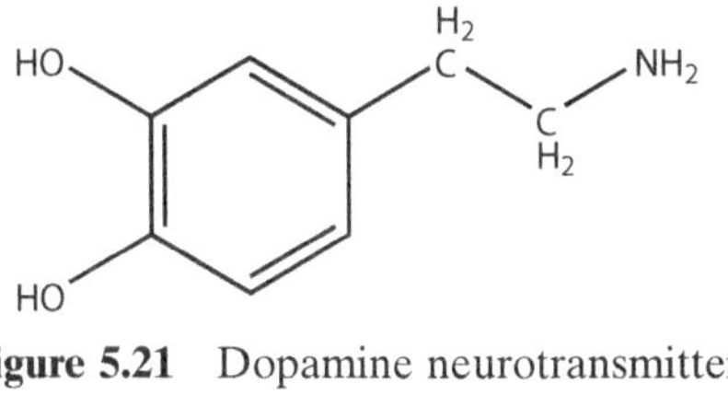

Figure 5.21 Dopamine neurotransmitter.

is a tertiary amine (see Fig. 5.18). Secondary and tertiary amines can take the form of rings like piperadine, shown in Figure 5.19. An amine can be thought of as ammonia (NH_3) with one or more of the hydrogen atoms replaced with alkyl groups. Amines are basic; they can accept a hydrogen ion (H^+) as shown in Figure 5.20. In addition, they are polar and can form a hydrogen bond. Amino acids, the building blocks of proteins, have an amino functional group. **Alkaloids** are biological amines produced by a variety of organisms, including many flowering plants. Some plants use alkaloids as a defense against insect attack. Many of these, including caffeine, cocaine, nicotine, psilocybin, and morphine are used by humans for their psychoactive effects. The psychoactivity of some alkaloids is due to their similarities to **neurotransmitters**, molecules (mostly amines) that carry signals between nerve cells. Figure 5.21 shows the neurotransmitter dopamine, which has amine and phenol functionality.

NAMING AMINES

Amines are named with the code for the longest chain containing a carbon that has a bond to the nitrogen atom, plus the suffix -anamine. If there are substituents on the amino group, their position is designated with *N*. If the amino group itself is considered to be part of a substituent, it is designated with the prefix amino-.

The carbon chain in Figure 5.22 is two atoms so the prefix is eth- and the suffix is -anamine. There are two one-carbon substituents (methyl groups) on the amino group: *N*,*N*-dimethylethanamine.

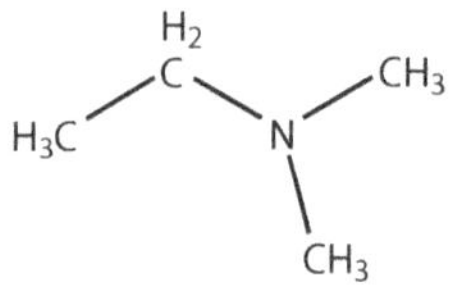

Figure 5.22 *N*,*N*-dimethylethylamine.

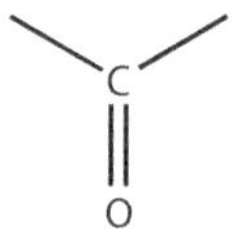

Figure 5.23 Carbonyl group.

Carbonyl Compounds

The **carbonyl** (CAR ben ill) group is C=O (Fig. 5.23). Carbon requires four bonds, so two additional bonds must be attached to the carbon atom. If these additional bonds are single bonds, then there are three electron groups on the carbon (a double bond and two single bonds); the geometry is trigonal planar. The atoms or groups attached to these bonds determine the type of compound. The carbonyl group is polar and can form hydrogen bonds with –OH or –NH groups.

Aldehydes and Ketones

If the carbonyl carbon has bonds to two carbon atoms, the compound is a **ketone** (KEY tone). **Acetone** (official name: propanone), an ingredient in nail polish remover, is the simplest ketone (Fig. 5.24). If the carbonyl carbon is attached to a carbon atom and a hydrogen atom or to two hydrogen atoms — that is, it is at the end of the carbon chain — the compound is an **aldehyde** (AL duh hide). Acetaldehyde (Fig. 5.24, officially ethanal) is a precursor to ethanol in the fermentation process. Aldehydes can undergo an addition reaction of the O–H of an alcohol across the C=O bond to give a **hemiacetal** (Fig. 5.25).

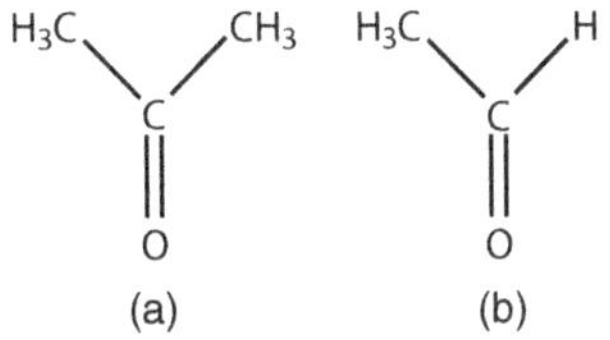

FUNCTIONAL GROUPS 101

Figure 5.24 Carbonyl compounds: (a) acetone and (b) acetaldehyde.

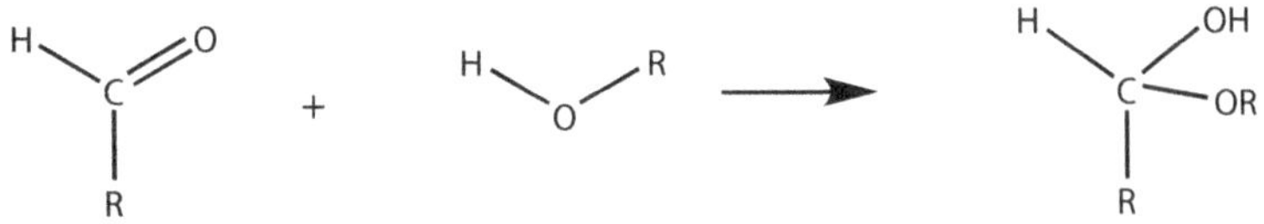

Figure 5.25 Addition to give hemiacetal.

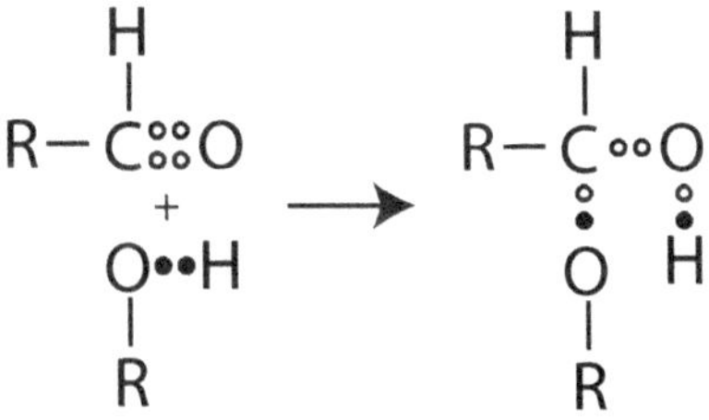

Figure 5.26 Hemiacetal electron pairs.

Ketones do the same to produce **hemiketals**. Figure 5.26 shows the details of the addition. You can identify hemiacetals or hemiketals because they have an –OH and an –OR group connected to the same carbon atom. Sugars that are important in brewing often exist as hemiacetals in the form of rings with the –OH group and the C=O groups both coming from the same molecule (see Fig. 6.3). Sugars can be thought of as starting out as aldehydes or ketones with –OH groups on the other carbons.

Keto-enol Tautomermism

An **enol** is a compound with an –OH group on a carbon that has a double bond to another carbon. Enols have a tendency to convert rapidly back and forth to ketones, a reaction called **tautomerism** (Fig. 5.27). Usually the molecules spend most of their time in the ketone form. But at any particular moment they can be in either form. If the double bond can be part of a system of alternating double and single bonds, the enol form can become the dominant form. An example is isohumulone, the principal bitter compound from hops. In this structure, shown in Figure 5.28, the related double bonds are marked in bold.

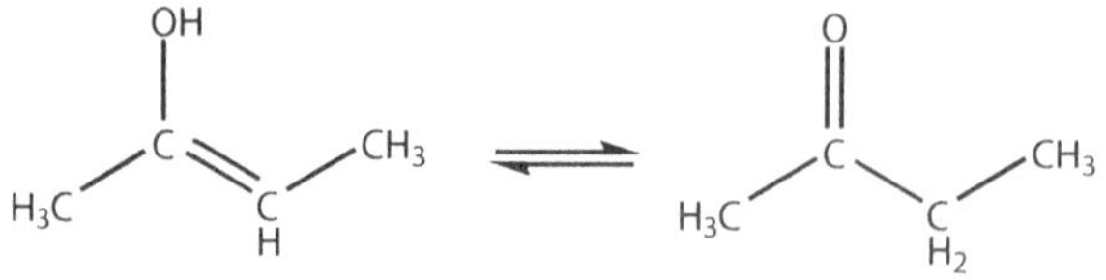

Figure 5.27 Keto-enol tautomerism.

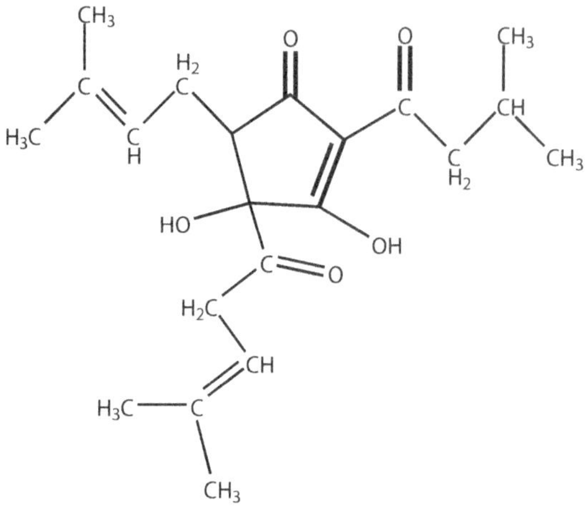

Figure 5.28 Isohumulone.

NAMING ALDEHYDES AND KETONES

Aldehydes are named with the suffix -anal. The C=O on an aldehyde is always at the end (carbon 1), so no number is needed. Common names for aldehydes are made from the common name of the corresponding carboxylic acid and replacing -ic acid with -aldehyde.

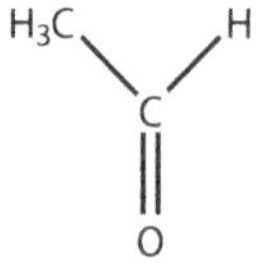

Figure 5.29 Acetaldehyde.

The molecule in Figure 5.29 has two carbons: ethanal. The common name, derived from acetic acid, is acetaldehyde.

The official name for a ketone uses the suffix -anone and a prefix indicating the position of the C=O carbon. Common names give the names of the alkyl groups attached to the C=O followed by ketone.

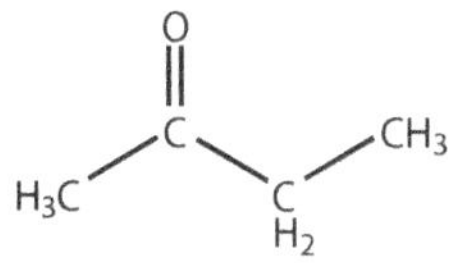

FUNCTIONAL GROUPS 103

$$\underset{H_3C}{}\overset{O}{\underset{}{\overset{\|}{C}}}\underset{H_2}{\overset{}{C}}CH_3$$

Figure 5.30 Butanone.

The molecule in Figure 5.30 has four carbon atoms: butanone. The common name is methylethyl ketone.

For aldehydes and ketones, if the double-bonded oxygen is a substituent, it is designated with the prefix oxo-.

Carboxylic Acids

If one group attached to the carbonyl carbon is an –OH group, the compound is a **carboxylic acid** (Fig. 5.31). The –OH group on a carboxylic acid can donate a hydrogen ion; it is an acid. The acidity of carboxylic acids is much higher than that of alcohols or phenols. They can lower the pH of water by a good deal.

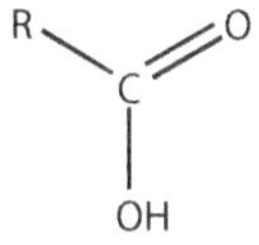

Figure 5.31 Carboxylic acid.

NAMING CARBOXYLIC ACIDS

Carboxylic acids take the suffix -anoic acid in the official naming system, but several of them have well-established common names. The acid group must be on the end of the chain (carbon 1) so no number is needed. If the –COOH group is a substituent; it gets the prefix carboxy-.

Carbons	Systematic Name	Common Name
1	Methanoic acid	Formic acid
2	Ethanoic acid	Acetic acid
3	Propanoic acid	Propionic acid
4	Butanoic acid	Butyric acid
6	Hexanoic acid	Caproic acid
8	Octanoic acid	Caprylic acid
10	Decanoic acid	Capric acid
18	Octadecanoic acid	Stearic acid

(continued)

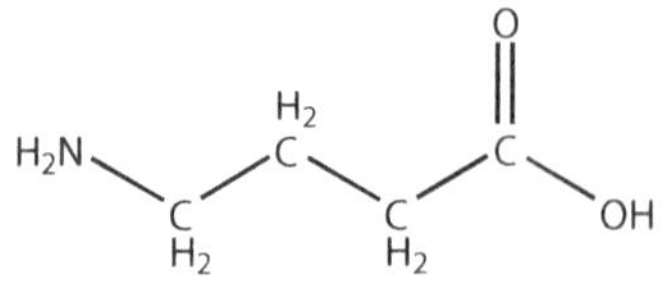

Figure 5.32 4-Aminobutanoic acid.

The molecule in Figure 5.32 has four carbons, so the stem is butanoic acid. There is an amine functionality on carbon 4 counting from the COOH end: 4-aminobutanoic acid.

Chains with two –COOH groups are officially named with the -dioic acid suffix, but they are usually known by common names.

Carbons	Systematic Name	Common Name
2	Ethanedioic acid	Oxalic acid
3	Propanedioic acid	Malonic acid
4	Butanedioic acid	Succinic acid
5	Pentanedioic acid	Glutaric acid
6	Hexanedioic acid	Adipic acid

We will mention two dicarboxylic acids with C=C bonds. Fumaric and maleic acids (Fig. 5.33), intermediates in cell respiration, are cis–trans isomers.

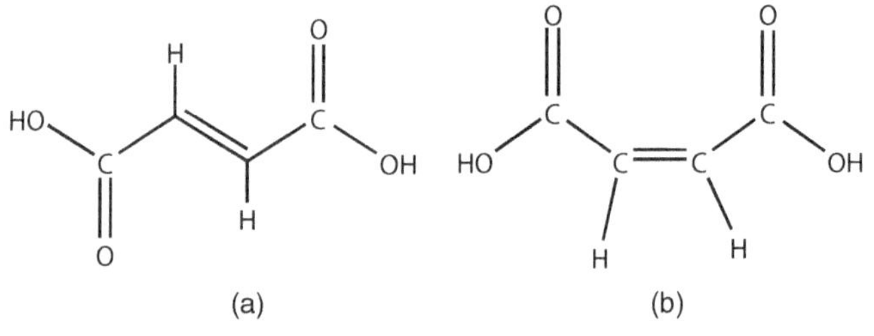

Figure 5.33 (a) Fumaric acid and (b) maleic acid.

Esters

Figure 5.34 shows how the –OH group of a carboxylic acid and that of an alcohol can link up by a dehydration reaction to form water and a product called an **ester**. An ester is a carbonyl compound in which one of the groups attached to the carbonyl carbon is –OR. Fats and oils are esters in which the

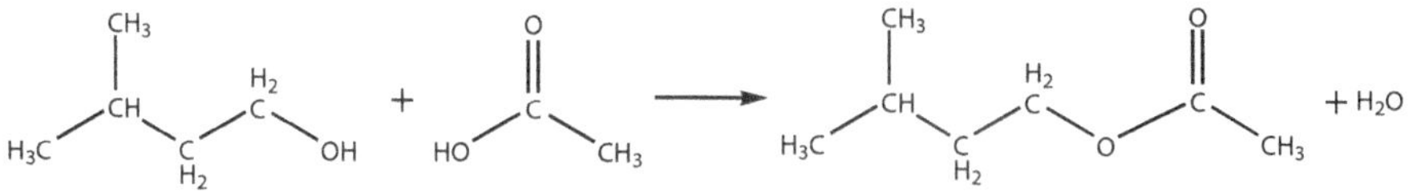

Figure 5.34 Ester formation: isoamyl alcohol + acetic acid → isoamyl acetate.

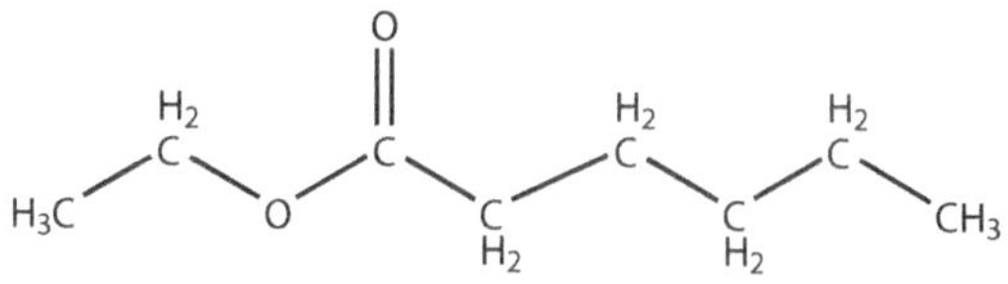

Figure 5.35 Ethyl hexanoate.

alcohol is glycerol, a compound with three –OH groups, and the acids, called fatty acids, have long chains, usually of 12 to 22 carbon atoms. Certain esters provide fruity aromas to beer, especially ale. For example, ethyl hexanoate, also called ethyl caproate (Fig. 5.35), is an ester made by condensing ethanol with a six-carbon carboxylic acid. It has the aroma of apples.

NAMING ESTERS

The name of an ester has two parts. The alkyl part, that is, the part of the molecule that came from the alcohol, is named first with the suffix -yl. The acyl part, which is the part that came from the acid, is named second with the -oic acid (or -ic acid) replaced with -ate. The official or the common name for the acid can be used.

In Figure 5.36 the group with C=O has two carbons: ethanoate (or acetate). The alkyl part is one carbon: methyl. The official name is methyl ethanoate. The common name is methyl acetate.

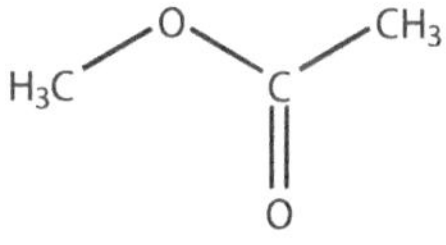

Figure 5.36 Methyl ethanoate.

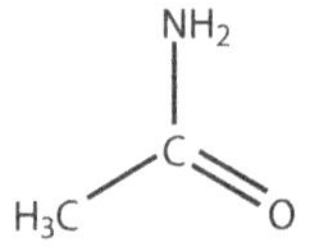

Figure 5.37 Acetamide.

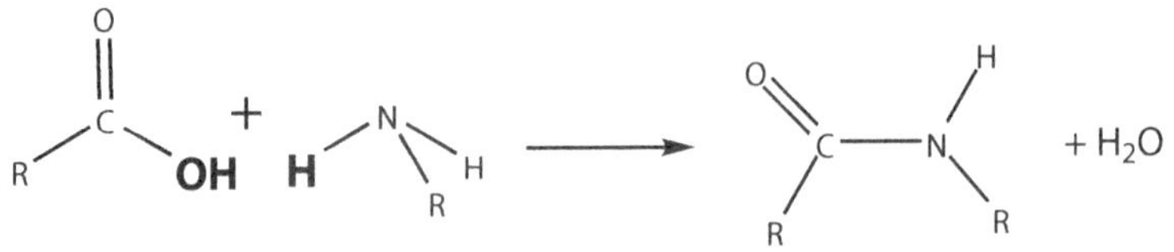

Figure 5.38 Amide formation.

Amides

If one group on the carbonyl carbon is an amino group (–NH_2) or a substituted amino group (–NR_2), the compound is an **amide** (AM id or AM ide) (Fig. 5.37). Amides are formed by the reaction of carboxylic acids with amines (Fig. 5.38). This is the reaction that links the parts of proteins into long chains. Because of an interaction between the C=O bond and the nitrogen unshared pair, the shape about the N atom is trigonal planar instead of the expected tripod. Compared to the amino group on an amine, the amino group on an amide is a very weak base.

NAMING AMIDES

The name of an amide is generated from that of the corresponding acid by changing -oic acid or -ic acid to -amide. If there are substituents on the amino group, they are located with an *N* prefix.

In Figure 5.39 the C=O is on a two carbon chain; the corresponding acid is ethanoic acid or acetic acid. The stem is ethanamide or acetamide. There is a methyl group on the nitrogen, so the official name is *N*-methylethanamide. The common name is *N*-methylacetamide.

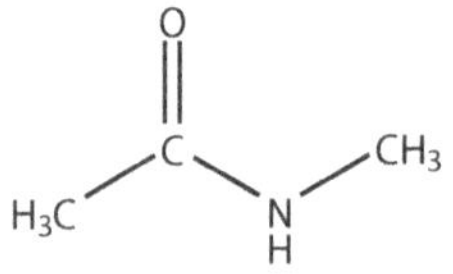

Figure 5.39 N-methylacetamide.

5.3 USING THE FUNCTIONAL GROUP GUIDE

The Functional Group Guide (Fig. 5.40 and Fig. 5.41) is a graphic decision tree that can help identify some compounds that are important in beer. We will follow an example to show how it works. Our task is to classify the compound shown below.

$$H_3C-\underset{H_2}{C}-O-\underset{H_2}{C}-OH$$

- Enter the Guide (Fig. 5.40) at "Start."
- The first decision is whether there are C–H or C–C bonds.
- The molecule has two C–C bonds and seven C–H bonds: the answer is yes.
- Following the line labeled "Yes" we get to the next decision box: Is there C and H only? This compound also has two O atoms, so the answer is no.
- The next box asks if there is a C=O. The answer is no, this leads to A, a link to Figure 5.41.
- We enter Figure 5.41 at A; the first box asks if there is an –OH, which there is on the right.
- The next decision is whether there is an O–C on the same C as the –OH; there is, on the left, so the answer is yes.
- The compound is a hemiacetal.

108 INTRODUCTION TO ORGANIC CHEMISTRY

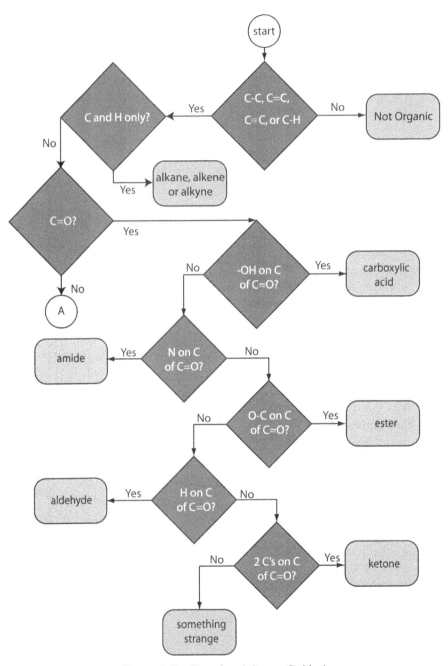

Figure 5.40 Functional Group Guide 1.

USING THE FUNCTIONAL GROUP GUIDE **109**

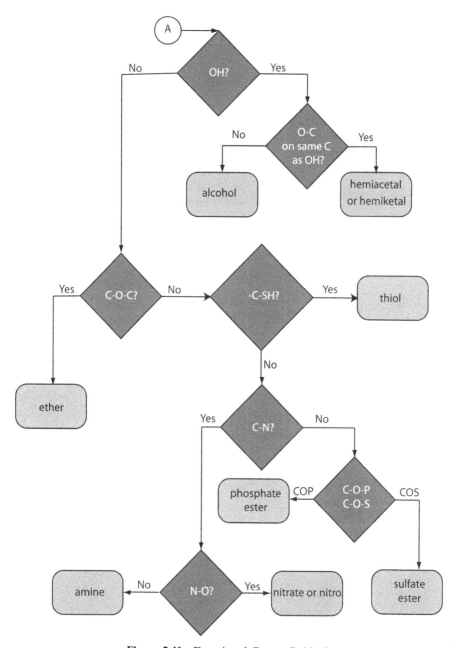

Figure 5.41 Functional Group Guide 2.

BIBLIOGRAPHY

Guinn, Denise; and Brewer, Rebecca. *Essentials of General, Organic and Biochemistry.* Freeman, **2009**. One version comes with a small model kit.

QUESTIONS

5.1. Draw the following as skeletal structures:

5.2. Draw the following as Lewis structures showing all atoms and bonds:

5.3. How does an aldehyde differ from a ketone?

5.4. What is an alcohol?

5.5. Classify each of the following compounds:

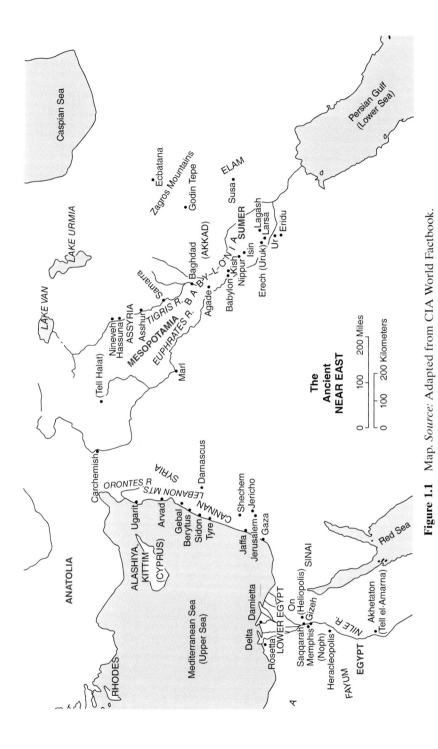

Figure 1.1 Map. *Source*: Adapted from CIA World Factbook.

The Chemistry of Beer: The Science in the Suds, First Edition. Roger Barth.
© 2013 John Wiley & Sons, Inc. Published 2013 by John Wiley & Sons, Inc.

Figure 1.2 A 2600 year old Sumerian–Akkadian dictionary of brewing terms, Metropolitan Museum of Art, New York.

Figure 1.3 Ancient Egyptian tile: grinding grain to packaging beer, Carlsberg–Israel Visitor Center, Ashkelon, Israel.

Figure 1.4 Hops on trellis.

Figure 2.2 Barley.

Figure 2.4 Female hop flowers and leaves.

Figure 2.10 Fermenter, home brewing.

Figure 3.6 Ionic bonding, sodium chloride crystal. Photo by Roger Barth.

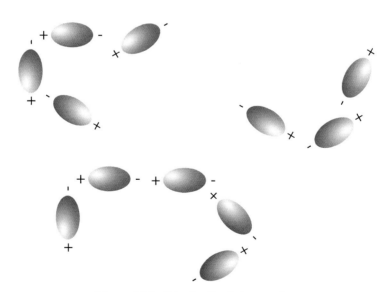

Figure 3.21 Dipole–dipole interactions.

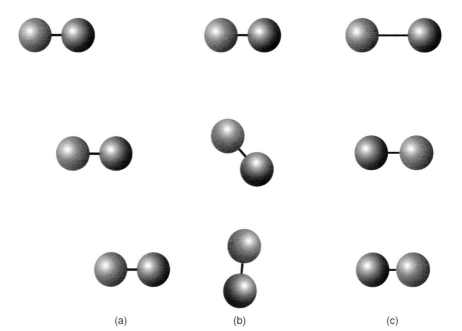

Figure 3.23 (a) Translation, (b) rotation, and (c) vibration.

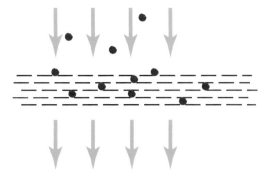

Figure 4.7 Filter.

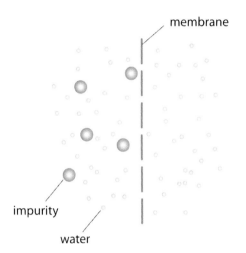

Figure 4.8 Osmosis.

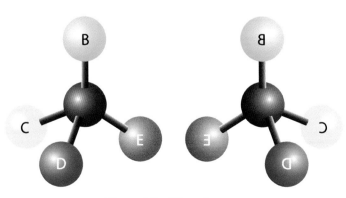

Figure 6.5 Mirror images.

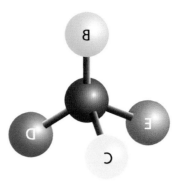

Figure 6.6 Rotated mirror image.

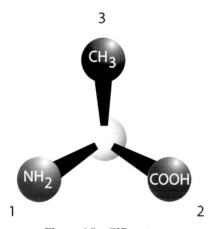

Figure 6.8 CIP system.

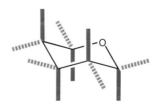

Figure 6.11 Axial–equatorial. Solid (blue): axial; dashed (orange): equatorial.

Figure 7.1 (a) Whole malt and (b) crushed malt.

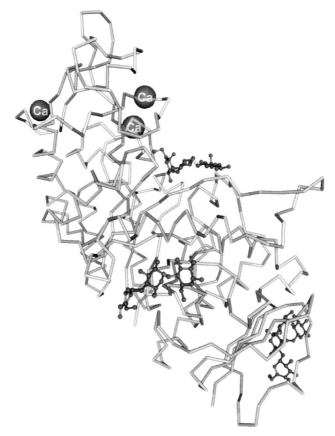

Figure 7.10 Barley alpha-amylase. (*Source:* From A. Kadziola et. al., *J. Mol. Biol.* **1994**, *239*: 104–121.)

Figure 8.1 Inside the lauter tun. Yuengling Beer Company, Pottsville, Pennsylvania.

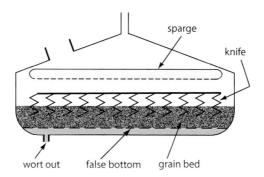

Figure 8.2 Lauter tun.

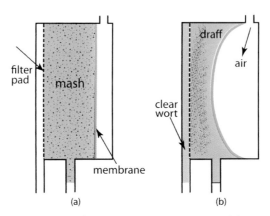

Figure 8.3 Mash filtration: (a) chamber with mash and (b) wort pressed out.

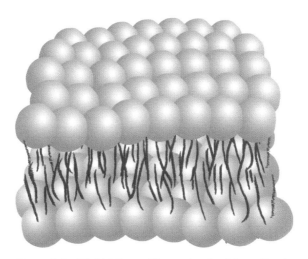

Figure 9.2 Lipid bilayer. Illustration by Marcy Barth.

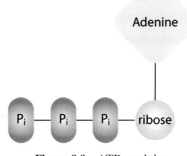

Figure 9.9 ATP modules.

Figure 10.1 Hydrometer.

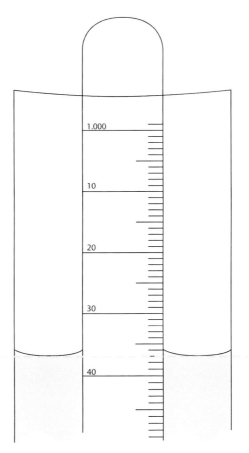

Figure 10.2 Reading a hydrometer.

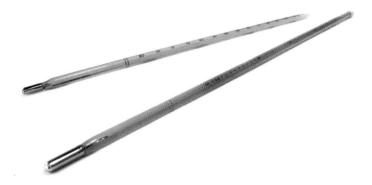

Figure 10.4 Liquid-in-glass thermometers.

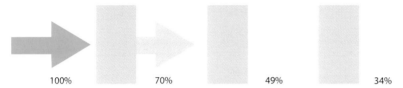

100%   70%   49%   34%

Figure 10.8 Light absorption.

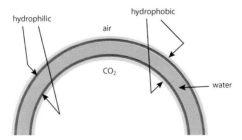

Figure 13.5 Surfactant on bubble.

Figure 16.11 Vorlauf.

Figure 16.12 Sparging.

(f) $H_3C-CH_2-O-C(=O)-CH_3$ ... (structures f, g, h shown)

5.6. Draw each structure in Question 5.5 as a skeleton structure.

5.7. For the structures in Question 5.5:

   a. Find the carbon atoms that are double-bonded to an oxygen.

   b. Find the carbon atoms that have a double bond to an oxygen and a single bond to another oxygen.

   c. Find an OH group attached to a carbon that also has a C=O bond.

   d. Find any molecules with two or more double bonds that alternate with single bonds.

5.8. What is the characteristic reaction of a double bond?

5.9. What functional groups join together to form an ester?

*__5.10.__ Draw the structure of the ester made from acetic acid and isobutanol.

(acetic acid and isobutanol structures shown)

*__5.11.__ Draw the structure of the amide made from acetic acid (see above) and 2-propylamine.

*__5.12.__ Draw the following compounds: butanamide, butan-2-ol, 3-methyl-1-pentanamine, 3-ethylhexanoic acid, and propyl pentanoate.

*__5.13.__ Name the structures shown in Question 5.5

Aldopyranoses, six-carbon aldose sugars in ring form.

CHAPTER 6

SUGARS AND STARCHES

The essence of brewing is to convert starch into sugar and the sugar into alcohol. The starch to sugar conversion takes place during mashing, and the sugar to alcohol conversion takes place during fermentation.

6.1 MONOSACCHARIDES

Sugar and starch are generic names for families of compounds that belong to a larger group called **carbohydrates**. Carbohydrates contain only carbon, hydrogen, and oxygen. A carbohydrate molecule always has exactly twice as many hydrogen atoms as oxygen atoms. The basic unit of a carbohydrate is an individual sugar molecule called a **monosaccharide** (Fig. 6.1). More complicated carbohydrates are made from two or more connected monosaccharide molecules. A monosaccharide has the same number of oxygen atoms as carbon atoms and, of course, twice as many hydrogen atoms, giving a general formula $C_nH_{2n}O_n$. A monosaccharide starts out as a chain of three or more carbon atoms with C=O on one of the carbons (an aldehyde or ketone) and an –OH group on each of the other carbons. If the sugar started as an aldehyde it is called an **aldose**; if it started as a ketone, it is a **ketose**. An example of a three-carbon aldose is glyceraldehyde (Fig. 6.2). This form of the sugar is called the open-chain form. Many five- and six-carbon sugars important in brewing exist mainly in a ring form. In the ring form of a sugar, one of the –OH groups on the open-chain form undergoes an addition reaction across the C=O bond in

The Chemistry of Beer: The Science in the Suds, First Edition. Roger Barth.
© 2013 John Wiley & Sons, Inc. Published 2013 by John Wiley & Sons, Inc.

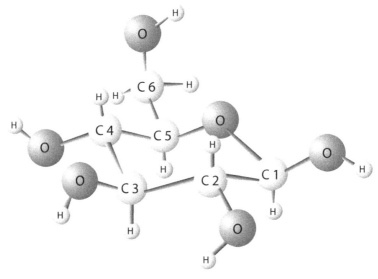

Figure 6.1 Beta-D-glucose (sugar).

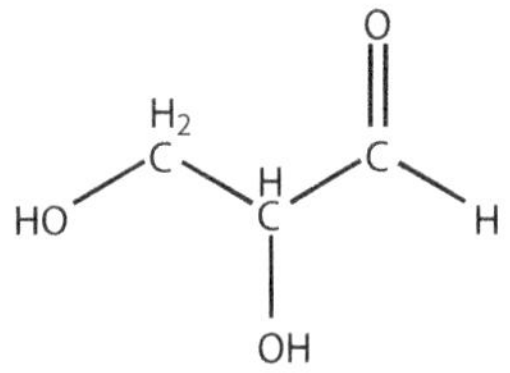

Figure 6.2 Glyceraldehyde.

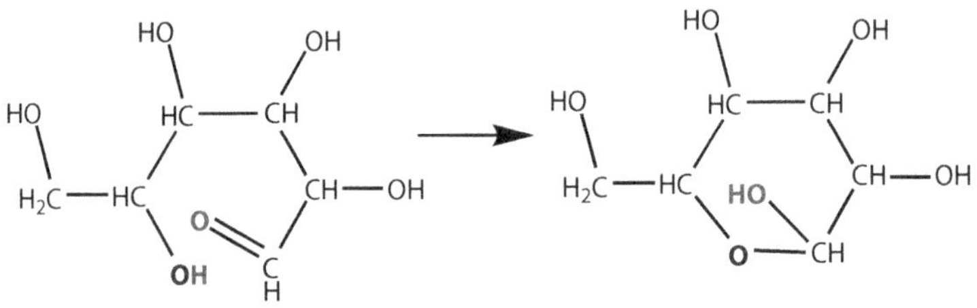

Figure 6.3 Hemiacetal formation.

the same molecule forming a hemiacetal or hemiketal in the form of a ring (Fig. 6.3). The reaction to form the ring is reversible in water solution, so a dissolved sugar molecule will spend a small fraction of its time in the open-chain form. The main sugars in brewing consist of rings of five or six atoms, one of which is oxygen and the others are carbon. Every carbon atom except one (C-5 in Fig. 6.1) has an –OH group attached. The one without the –OH is

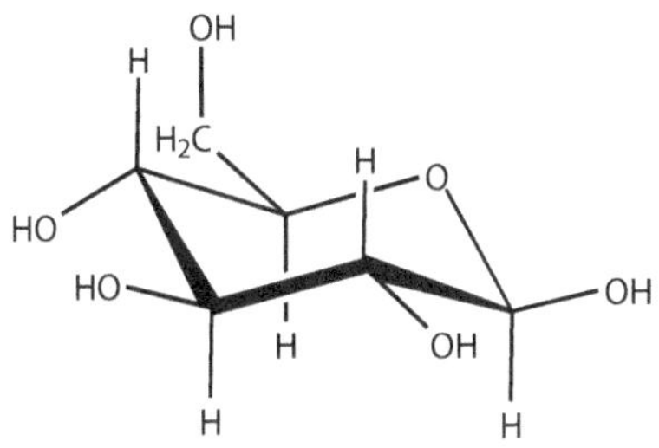

Figure 6.4 Beta-D-glucose (skeleton form).

attached to the ring oxygen. The carbon that is attached on the other side of the ring oxygen (C-1) has an –OH, making it the only carbon attached to two oxygen atoms. This is the carbon that originated as the carbonyl (C=O) carbon; it is called the **anomeric** carbon.

The most important monosaccharide involved in beer is **glucose**, shown in Figure 6.1 and in skeleton form in Figure 6.4. There are six carbon atoms in glucose. Of these, all but one are attached to four different things. C-1 to C-4 are each attached to two ring neighbors, to one hydrogen atom, and to one –OH group. In the case of C-2 to C-4, even though the ring neighbors are both carbon atoms, they are different from one another in terms of their positions in the ring and the atoms attached to them. C-5 is attached to two ring neighbors, to C-6, and to a hydrogen atom. A carbon atom in any class of compound that is attached to four different things is called an **asymmetric carbon** atom. The last carbon (C-6) in glucose is attached to C-5, to an –OH group, and to two hydrogen atoms. Because the two hydrogen atoms are identical, C-6 is not asymmetric. Carbons atoms with multiple bonds are never asymmetric.

6.2 CHIRALITY

Asymmetric carbon atoms have a surprising property: there are two ways to put them together. We can think of these as a right-handed and a left-handed version. Molecules that have a different right- and left-handed version are said to be **chiral** (KYE-r'l) (Greek: *kheir*, hand). A molecule with one asymmetric carbon exists in two forms that are mirror images of one another. Figure 6.5 shows a molecule with an asymmetric carbon and its mirror image. Figure 6.6 shows the molecule on the right in Figure 6.5 rotated to bring ball E into the position it occupied before reflection as shown on the left in Figure 6.5. You can see that, although the rotation makes balls B and E match up, balls C and D do not match up. No amount of rotating will make all the balls match up. The molecule on the left in Figure 6.5 and its mirror image on the right can only be made to match up by breaking and moving chemical bonds. So the molecules in Figure 6.5 are different compounds. A nonchiral molecule, by

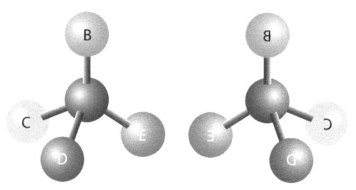

Figure 6.5 Mirror images. (See color insert.)

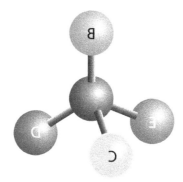

Figure 6.6 Rotated mirror image. (See color insert.)

contrast, has a mirror image, but the mirror image is identical to the original molecule.

Chiral compounds that are mirror images of one another are called **enantiomers** (Greek: *enantios*, opposite + *meros*, part). Enantiomers have identical properties to one another except for interactions with things that have handedness, including polarized light and other chiral molecules, such as those made in living cells. Pairs of enantiomers are usually given the same name and distinguished by prefixes D- or L- (Latin: *dexter*, right-handed; and French: *lévo-*, left) or, more systematically, (*R*)- or (*S*)- (see box on "Absolute Configurations"). Chirality occurs in many classes of compounds important for beer, including carbohydrates, amino acids, proteins, sterols, hop bitter compounds, and many others. In nearly every case, only one of the two enantiomers occurs naturally.

It is important to remember that a carbon with a double or triple bond is not attached to four things, so it can't be asymmetric. Any implied (not shown) hydrogen atoms count as something attached to a carbon atom.

ABSOLUTE CONFIGURATIONS: A CLOSER LOOK

The usual D- and L- prefixes for naming enantiomers have historical origins and are now strictly conventional. The absolute configuration about each asymmetric carbon is specified with a logical system called the Cahn–Ingold–Prelog (CIP) system. Here is the process.

- Assign to each of the four items attached to the asymmetric carbon a priority according to the rules below, the first being the highest priority, and the fourth being the lowest.
- Arrange the molecule so that the fourth priority group is pointing directly away from you.
- Look at the arrangement of the remaining three items from first to third priority.
- If this arrangement is clockwise, the configuration is *R* (Latin: *rectus*, right).
- If the arrangement is counterclockwise, the configuration is *S* (Latin: *sinister*, left).

The first priority is assigned to the group whose first attached atom has the highest atomic number. To break ties, go progressively further (more bonds) and compare atomic numbers of the atoms in order of decreasing atomic number until the tie is broken.

On the alanine molecule in Figure 6.7, the four items attached to the asymmetric carbon are hydrogen (atomic number = 1), carbon (6), carbon (6), and nitrogen (7). One of the carbons is –CH$_3$; the atoms attached to it are hydrogen with atomic number 1, the other carbon is –COOH; the atoms attached to it are oxygen with atomic number 8. The priorities are –NH$_2$ (first), –COOH (second), –CH$_3$ (third), and –H (fourth).

Figure 6.8 shows the alanine molecule rotated so the H atom (fourth priority) points directly away from us. We see that the first to third

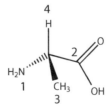

Figure 6.7 L-Alanine.

(*continued*)

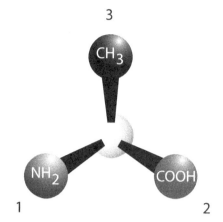

Figure 6.8 CIP system. (See color insert.)

priorities are arranged in a counterclockwise manner. The absolute configuration about the asymmetric carbon on L-alanine is *S*. This is a good deal easier to understand with models. Official chemical naming now uses (*R*)- and (*S*)- instead of D- and L- as prefixes. The official name for L-alanine is (*S*)-2-aminopropanoic acid.

Diastereomers

What if a compound has two asymmetric carbon atoms? Each of these can have one of two configurations, which we can designate *R* and *S*. So there are four different ways that the asymmetric carbons can be combined in the molecule: *RR*, *SS*, *RS*, and *SR*. The mirror image of a molecule has the configuration of each asymmetric carbon inverted; that is, each *R* becomes an *S* and each *S* becomes an *R*. On reflection through a mirror, *RR* becomes *SS*; they are enantiomers. The mirror image of *RS* is *SR*, so these two are enantiomers. But *RR* and *RS* are not enantiomers. They have different chemical and physical properties. It is worth noting that if the two sides of the molecule are identical, then *RS* and *SR* are identical so the molecule is not chiral, but it is still different from *RR* and *SS*. Compounds that differ by the configuration about one or more asymmetric carbons but are not mirror images are called **diastereomers**. Enantiomers and diastereomers are examples of **stereoisomers**, molecules that have the same atoms bonded in the same order, but with different arrangements in space. Figure 6.9 shows four forms of the four-carbon aldoses, erythrose and threose, in which the second and third carbon atoms are asymmetric. There are four stereoisomers forming two pairs of enantiomers. Erythrose has the –OH groups on asymmetric carbons going in opposite directions, one forward, one back. Threose has the two –OH groups going the same direction, either both forward (L-threose) or both back (D-threose).

CHIRALITY 119

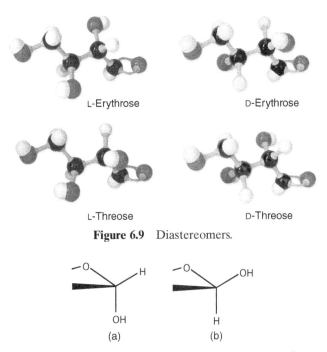

Figure 6.10 Anomeric carbon configuration: (a) alpha configuration and (b) beta configuration.

Glucose in its ring form has five asymmetric carbons, giving it a total of 32 stereoisomers forming 16 pairs of enantiomers. It turns out that the configuration on the anomeric carbon (the one attached to two oxygen atoms) is easily inverted because the ring can open and close at that point, so there are only four carbons whose configurations are fixed. This gives 16 compounds, or 8 pairs of enantiomers. The D enantiomer of each of these eight sugars, called allose, altrose, galactose, glucose, gulose, idose, mannose, and talose are shown on the facing page of the chapter opener. Galactose, glucose, and mannose are common in nature. We use Greek letters alpha (α) and beta (β) as prefixes when we need to distinguish the configuration on the anomeric carbon (see Fig. 6.10). Glucose is the main sugar used for energy in living organisms and it will be our focus.

Axial and Equatorial

A regular hexagon has an interior angle of 120 degrees. Because the bond angles on the carbon and the oxygen atoms in the six-atom ring of a glucose molecule are about 109.45 degrees, the rings in these sugars do not form flat hexagons. Instead, the ring is puckered. Each ring carbon has two of its four bonds going to its ring neighbors; the other two stick out of the ring. Because of the ring puckering, one of the bonds that stick out of the ring from each

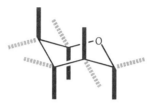

Figure 6.11 Axial–equatorial. Solid (blue): axial; dashed (orange): equatorial. (See color insert.)

carbon points close to the plane of the ring, like the spokes of a wheel (shown dashed in Fig. 6.11), the **equatorial** position. The other bond that sticks out of the ring points up or down perpendicular to the plane of the ring (parallel to the axle, if it had one), the **axial** position. Whether a particular atom or group occupies the axial or equatorial position depends on the configuration (handedness) about the ring carbon. The equatorial position is preferred for groups that are larger or have a higher concentration of electrons. In beta glucose, all the –OH groups and the –CH$_2$OH (C-6 on Fig. 6.10) are in the equatorial positions and the small hydrogen atoms occupy the more crowded axial positions. This makes beta-glucose particularly stable.

6.3 DISACCHARIDES

The –OH group on the anomeric carbon is more reactive than other –OH groups. This group can undergo a condensation reaction with an –OH group on another sugar molecule binding the sugar molecules together and releasing water:

$$G-OH + HO-G \rightarrow G-O-G + H_2O$$

The product is a two-sugar unit called a **disaccharide**. The –O– link connecting the anomeric carbon on a sugar molecule to another group is called a **glycosidic link**. Compounds made of glucose units connected by glycosidic links are called **glucans**. When the anomeric carbon gets tied up in a glycosidic link, the ring can no longer open, so the alpha or beta configuration becomes fixed, making these two forms separate compounds that do not readily interchange. If an axial (alpha) –OH on C-1 of one glucose reacts with –OH on C-4 (which is directly across the ring) of another glucose, the resulting disaccharide is **maltose**. When the original C-1 –OH group was in the alpha (axial) configuration, the connection between the glucose units is called an alpha-1→4 linkage, because it connects C-1 in the alpha configuration of one glucose to C-4 of the other. As shown in Figure 6.12, the maltose molecule is bent.

Cellobiose is the corresponding disaccharide in which the –OH on C-1 was in the beta (equatorial) position. Cellobiose has a beta-1→4 linkage. Cellobiose (Fig. 6.13) is a straight molecule in contrast to maltose, which is bent.

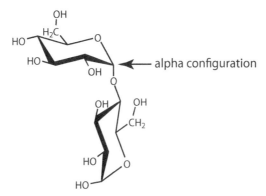

Figure 6.12 Maltose.

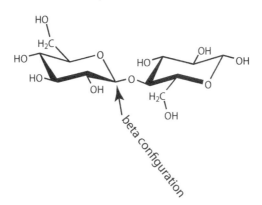

Figure 6.13 Cellobiose.

6.4 POLYSACCHARIDES

Both maltose and cellobiose can be extended. That is, we can make a glycosidic link from C-4 of another glucose to the C-1 of the right-hand glucose in the disaccharide, and we can make a link C-4 of the left-hand unit in the disaccharide to C-1 of another glucose. We can keep doing this until we have strung together hundreds or thousands of glucose units. If, as in maltose, the linkages are alpha-1→4, the resulting material is a **starch** called **amylose**. Because the maltose units are bent, the chains of rings in amylose (Fig. 6.14) are twisted.

If the linkages are beta-1→4, as in cellobiose, the resulting material is **cellulose**, a structural element in plants. Cellobiose is straight, so cellulose consists of straight chains of rings that easily form hydrogen bonds with adjacent chains giving stiff fibers. Cellulose is used by plants to provide stiffness; it is what makes wood woody. Cellulose is difficult to hydrolyze to sugars. Some fungi (but not yeast) and bacteria can do this. The resistance of cellulose to

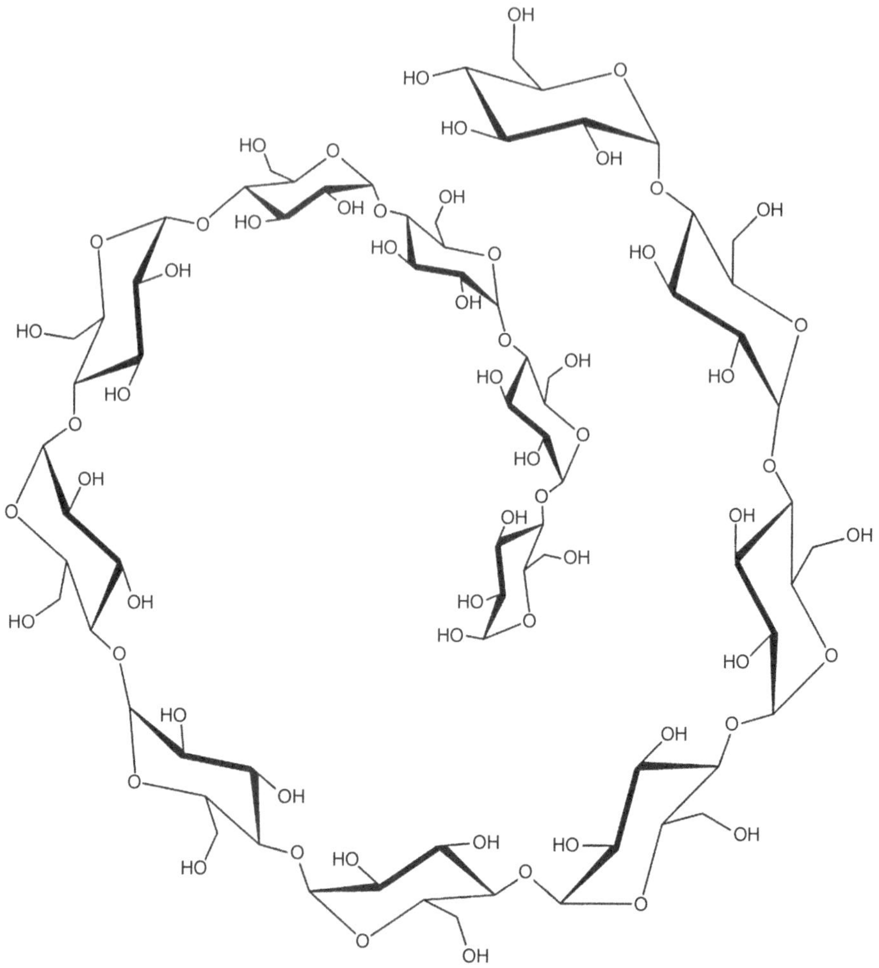

Figure 6.14 Amylose.

hydrolysis makes it possible for trees and humans to use wood as a long-lived structural material.

Because C-1 and C-4 are straight across the ring from one another, 1→4 linkages can form long, fairly organized chains. But an –OH group on C-1 can also react with the –OH groups on carbons other than C-4. A molecule of amylose, that is, a chain of glucose units with alpha-1→4 linkages, has free –OH groups remaining on C-2, C-3, and C-6. The C-1 hydroxide of a new glucose can link to one of these, usually to the –OH group of C-6, making a branch in the chain (Fig. 6.15). **Amylopectin** (Fig. 6.16), a starch found in plants, has branches about every 30 to 50 glucose units. Barley starch is typically about 25% amylose and 75% amylopectin. A strand of amylopectin has many free

KNOW YOUR CARBOHYDRATES

Carbohydrates can take a number of seemingly disparate forms. The situation is complicated because substances derived from and structurally related to carbohydrates are regarded in some contexts as carbohydrates. We will regard such materials as distinct from true carbohydrates.

1. A carbohydrate has only carbon, hydrogen, and oxygen. If any other element is present, it is not a carbohydrate.
2. Carbohydrates have two atoms of hydrogen for each atom of oxygen.
3. Every carbon atom is bonded to at least one oxygen atom.
4. Each unit of a carbohydrate is an aldehyde, ketone, hemiacetal, or acetal. Each carbon atom outside of those functionalities has an –OH group

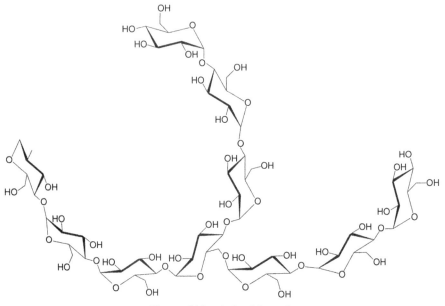

Figure 6.15 A 1→6 branch.

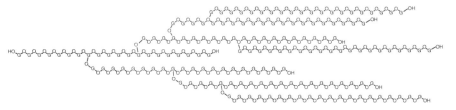

Figure 6.16 Amylopectin schematic.

ends, so reactions that break small groups of glucose molecules off from the chain ends are more rapid on amylopectin than on amylose. **Glycogen** has branches every 10–12 glucose units. Glycogen is a starch used to store energy by animals and fungi including yeast.

Gums

Gums are polysaccharides that dissolve to some extent in water. Gums can greatly increase the **viscosity** (flow resistance) of the water in which they dissolve, and many are used for this purpose. The gum of greatest interest to brewing comes from degradation of cell walls of grain hulls. This gum is known as beta-glucan (β-glucan) gum. Beta-glucan gum has beta-1→4 linkages like cellulose, but one in every four to six glucose units is connected with a beta-1→3 linkage as shown in Figure 6.17. This introduces a bend in the molecule so the chains do not fit together well. As a result, they make hydrogen bonds to water rather than to one another, which is why the gum dissolves.

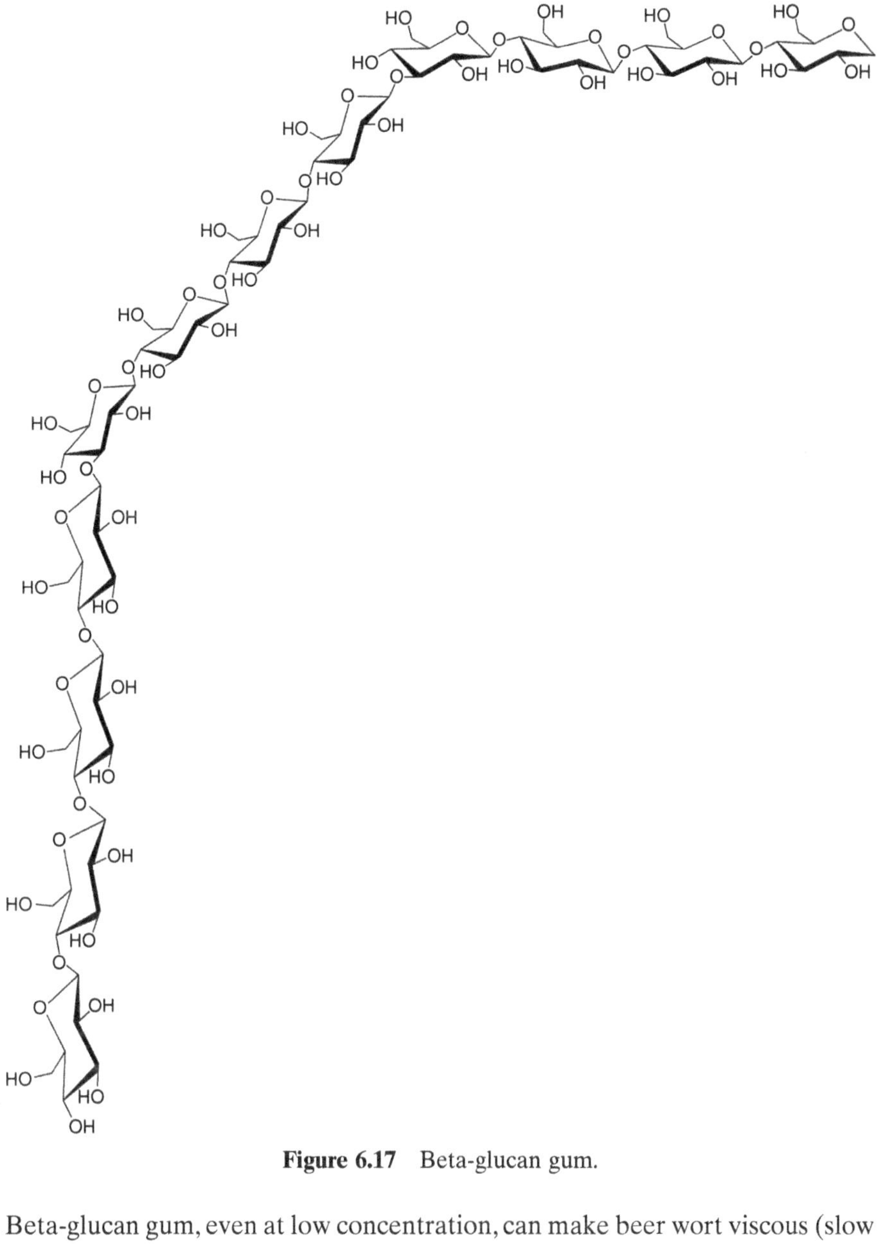

Figure 6.17 Beta-glucan gum.

Beta-glucan gum, even at low concentration, can make beer wort viscous (slow flowing), which makes the wort difficult to filter and pump. Beta-glucan gums can get into the wort from incompletely sprouted (undermodified) barley, from mashing in the presence of fine particles of barley hull, or from sparging for too long or with water that is too hot or at a high pH. The result can be a set mash, a condition in which the wort is nearly a gel and it runs very slowly if at all. Imagine a lauter tun with 300 barrels of wort that will not run out. There are other gums that have similar effects.

TESTING CARBOHYDRATES

Reducing Sugars and Tollens's Test

A sugar with an –OH group on the anomeric carbon that is not involved in a glycosidic link is called a **reducing sugar**. All of the sugars shown on the chapter opening facing page are reducing sugars, as are maltose (Fig. 6.12) and cellobiose (Fig. 6.13). Sucrose (see Question 6.12) is not a reducing sugar because the anomeric carbons of the glucose and fructose portions are linked to one another in a way that leaves no free –OH on an anomeric carbon. Starches and cellulose are not reducing sugars because they have only a single free anomeric carbon for hundreds or thousands of sugar units.

When any aldehyde or a ketone with an –OH group on a carbon adjacent to the carbonyl group is treated with a basic solution of the diammine silver ion, called **Tollens's reagent**, the silver ion is reduced to the metal, which can leave a mirror on the inside of glassware. Reducing sugars react in this way. This reaction, using glucose as the aldehyde, was formerly used to make mirrors on plate glass. The reaction is

$$R-CHO + 2OH^- + 2Ag(NH_3)_2^+ \rightarrow R-COOH + H_2O + 2NH_3 + 2Ag$$

Similar tests using copper(II) ion (Cu^{2+}) instead of silver, called Benedict's reagent and Fehling's reagent, provide the same information, but without the striking silver mirror.

Iodine Test

Amylose and amylopectin give a deep purple color when treated with dissolved iodine (the actual reactant is I_3^-). The color arises from the entry of I_3^- ions into the coils of the starch molecules. Cellulose, which has no coils, does not give this response. Counterfeit-detecting pens use this reaction to test paper money for starch, a common component of paper. Legitimate currency (as well as some counterfeit) is made without starch. Brewers often use the iodine test to determine if the starch in the grist has been fully hydrolyzed in mashing. If a sample from the mash gives a purple color, hydrolysis is not complete.

BIBLIOGRAPHY

Briggs, Dennis E.; Boulton, Chris A.; Brookes, Peter A.; and Stevens, Roger, *Brewing Science and Practice*. CRC Press, 2004, pp. 122–142. Review of carbohydrates in beer wort. Uses diagrams that follow an older convention.

QUESTIONS

6.1. Explain the following terms: monosaccharide, starch, amylopectin, anomeric carbon, chirality, and asymmetric carbon.

6.2. What is the difference between glucose and its diastereoisomers, like mannose?

6.3. Why is maltose bent, in contrast to cellobiose, which is straight?

6.4. What is the difference between amylose and amylopectin?

6.5. What is the difference between D-glucose and L-glucose?

6.6. How does starch differ from cellulose?

6.7. What is a gum? What effect can gums have on brewing?

6.8. Among the molecules below, tell which are carbohydrates.

128 SUGARS AND STARCHES

(f)

(g)

(h)

6.9. Mark all asymmetric carbons on the molecules shown in Question 6.8.

6.10. Identify the anomeric carbons on any carbohydrates shown in Question 6.8.

6.11. Mark the glycosidic link on any carbohydrates among the molecules in Question 6.8.

*6.12. Draw the monosaccharides that would result from the hydrolysis of sucrose.

Sucrose

Grist mill. Yuengling Beer Company, Pottsville, Pennsylvania.

CHAPTER 7

MILLING AND MASHING

The carbohydrate in grain is produced by the plant to nourish the baby plant, called an **embryo**. The carbohydrate in malted barley is only 15–20% sugars; the rest is starch. Starch is good for storage but it can't be used directly either by the embryo or by the yeast that performs the fermentation. Starch consists of long chains of **glucose** rings; it is insoluble in water (and beer) and can't be used by the yeasts that make beer. Yeast can use glucose (1 glucose unit), **maltose** (2 glucose units), **maltotriose** (3 glucose units), and a few others. Most of the carbohydrate from malt will only be available to the yeast after the glycosidic bonds holding the glucose rings together are hydrolyzed, turning the starch into sugar.

Under ordinary conditions, the hydrolysis reaction is immeasurably slow. As the seed sprouts, special proteins called **enzymes** are produced that greatly speed up the hydrolysis reaction by getting the reactants (starch and water) and the energy lined up in the right way to allow the reaction to occur quickly. Enzymes play the role of **catalysts**, they speed up reactions but they are not reactants or products. Living cells can maintain control over their reactions by controlling the amounts of enzymes or by modifying their activities. In a sprouting seed the enzymes convert starch to sugar over a period of days or weeks to meet the needs of the embryo until it can use energy from sunlight. The brewer can't wait that long so the first two steps of brewing—**milling** and **mashing**—are designed to convert the starch to sugar in about an hour.

The Chemistry of Beer: The Science in the Suds, First Edition. Roger Barth.
© 2013 John Wiley & Sons, Inc. Published 2013 by John Wiley & Sons, Inc.

132 MILLING AND MASHING

(a)   (b)

Figure 7.1 (a) Whole malt and (b) crushed malt. (See color insert.)

7.1 MILLING

To make beer, the brewer must get the solid starch in grain to react with liquid water to produce sugar. When a reaction involves two phases of matter, solid and liquid in this case, the reaction can only occur at the interface where the phases meet. The reaction goes faster if the solid grain is broken into smaller pieces, because breaking up the starch particles exposes additional surface area at which the reactants meet (Fig. 7.1). The grains are broken up in a device called a **mill**. The usual type of mill crushes the grains between steel rollers. The particle size of the crushed grain, called grist, can be controlled to some extent by the spacing between the rollers, called the gap. Another type of mill is the hammer mill. A hammer mill has a rotor attached to swinging bars of metal, called hammers. The rotor spins rapidly causing the hammers to smash the grain to flour.

In most breweries, the grain bed serves as a filter to clarify the sugary liquid, called **wort**; the wort has to flow through a bed of **grist** (crushed grain). If the grist particles are large, the spaces between them will be large, and the wort will flow easily. Contact between the large starch particles and the water is minimal, leading to less sugar for a certain amount of grain, hence more grain for a certain amount of beer. If the particles are very small, starch to sugar conversion is good, but the spaces between the particles are small, making it difficult for the wort to flow through the grain bed. Between these extremes there is a certain size that makes the best compromise for a particular brewing operation. Some breweries separate the grains from the wort by filtration under pressure. In this case, the grain can be reduced to flour with a hammer mill, enhancing the yield of sugar. The pressure in the filter unit makes up for the increased flow resistance.

When grain is crushed, or even when it is moved around, fine dust rubs off. Because of its large surface area, this dust can burn or even explode from sparks or other sources of ignition. The tiny dust particles don't settle easily, so the dust tends to form clouds that get everywhere. In addition to the danger

Figure 7.2 Mashing in a water cooler.

of fire and explosion, the dust particles carry bacteria that can have a bad effect on the beer. The dust is also unhealthy to breathe. For these reasons, the milling operation has to be separated from the other brewing processes. In most cases the air from the milling area must be filtered to remove the dust before it can be released. Some equipment allows the grain to be moistened before milling, eliminating the dust problem.

7.2 MASHING

The second step of brewing is **mashing**, which is soaking the grist in hot water to allow the water and enzymes to act on the starch (Fig. 7.2). The purpose of mashing is to hydrolyze as much of the starch as possible to sugars that the yeast can convert into ethanol and carbon dioxide. If the water is not too alkaline, components of the malt will usually bring the mash pH to a value in the range of 5.2–5.8. Values of the pH outside this range can affect the functioning of the mashing enzymes. High pH can cause undesired compounds to leach from the grain hulls into the beer.

Three things have to happen during mashing: **gelatinization**, **liquefication**, and **saccharification**.

Gelatinization

The starch granules in grain consist of tightly packed chains of glucose rings. The chains stick to one another because of hydrogen bonding between the

–OH groups of rings on adjacent chains. Water has to work its way into the granules and get between the chains so that they can be separated. This causes the –OH groups on the starch molecules to form hydrogen bonds to the –OH groups on water molecules instead of to those on the other starch molecules. As long as the starch chains are tightly packed together, the enzymes and the water are not able to make good contact with them to chop them into sugars. The movement of water into the starch granules to separate the starch chains is called gelatinization. Hotter water causes faster gelatinization. Some grains are easier to gelatinize than others. Rye, barley, wheat, and oats gelatinize completely below 70 °C (158 °F). Certain grains like rice, maize (corn), sorghum, and millet have to be cooked before mashing because they do not gelatinize adequately under mashing conditions.

Liquefication

There are two ways to break up starch molecules. One way is to chop up the chains at various points in the middle; the other is to nibble bits off the ends. In both cases, the chemical reaction is enzyme-facilitated hydrolysis of glycosidic links. Chopping in the middle is the mode of action of **alpha-amylase** (α-amylase), one of the two main starch-degrading enzymes in the mash. Nibbling on the end is the mode of the other enzyme, **beta-amylase** (β-amylase). Chopping starches in the middle gives shorter, more soluble chains of glucose rings called **dextrins**. Cutting starch molecules to give soluble dextrins is called liquefication. Liquefication brings carbohydrates into solution; that is, it increases the fraction of the grain that dissolves. This fraction is called the **extract** in brewing jargon. Regrettably, the word "extract" has other beer-related meanings (see Glossary). High extract is desirable because it means more beer from a given amount of grain.

Saccharification

Yeast can only use very short chains of sugar rings. They do best with one to three sugar units. The action of beta-amylase (β-amylase) is to nibble off two rings from the end of a starch chain or dextrin. Two glucose rings form a sugar called **maltose**. If the starch is well liquefied, there will be plenty of chain ends for β-amylase to work on. This end-nibbling reaction is called **saccharification** (Latin: *saccharum*, sugar) because it turns starches into sugars that the yeast can use. The process increases the **fermentability** of the mixture, that is, the extent to which the carbohydrates can be used by the yeast to make alcohol and carbon dioxide. The desired degree of fermentability depends on the style of beer to be produced. A hearty English ale might be expected to have a good amount of "body" provided by unfermented carbohydrates. An American light beer would hardly be more than alcohol and water.

7.3 ENZYMES AND PROTEINS

Chemical reactions usually involve breaking bonds and making new bonds. For example, breaking a glycosidic link in a starch molecule involves breaking an O–C bond between two joined rings and an O–H bond in water, then making two new O–H bonds. This reaction, adding water and breaking the chain, is called **hydrolysis** (Greek: *hydor-*, water + *lysis*, loosening) (Fig. 7.3).

Catalysts

Before the new bonds can be made, the old bonds must be broken. It takes energy to pull apart the atoms in a bond, so before a chemical reaction can occur, energy must be provided to break the bonds. This amount of energy is called the **activation energy**. Figure 7.4 shows the potential energy as the reaction progresses for a simple molecular encounter. The highest point on the potential energy curve represents the **transition state**. At the transition state, the bond on the left is half broken and the bond on the right is half made. After the transition state, it is downhill all the way. The amount of energy needed to reach the transition state is the activation energy, marked E_a. If the energy available is not enough to push the reaction up the hill to the transition state, the reaction fails. Another requirement for a reaction is that the groups that are going to react need to be near enough and in the right positions to exchange the atoms connected by the chemical bonds. If we just mix everything together and wait for the right conditions, it could take a very long time. In many cases some other substance, not the reactants or products, can get involved and make the reaction faster. Such a substance is called a **catalyst**. The catalyst may do its job by helping line up the groups that need to react, or by funneling energy, electrons, or hydrogen ions to the right place. An example from the mechanical world is the catalytic converter in a car. Tiny bits of platinum and cerium oxide get the oxygen molecules into the right place to react with carbon monoxide and other pollutants from the engine under conditions where these reactions would otherwise be very slow.

$$2CO + O_2 \rightarrow 2CO_2$$

The catalysts in living organisms are called **enzymes**. Enzymes are made from proteins, sometimes with metal ions, water molecules, or other ions or molecules attached.

Amino Acids

Proteins, like starches, are long chains (**polymers**, in chemistry jargon) made up of smaller subunits. Unlike starches, in which all the subunits are the same type of sugar, there are 22 different subunits that are specified by the genetic code and occur in proteins. These subunits, called amino acids, are shown in

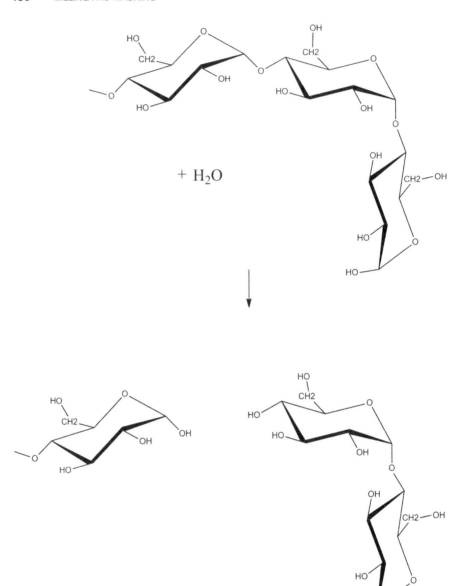

Figure 7.3 Hydrolysis of starch.

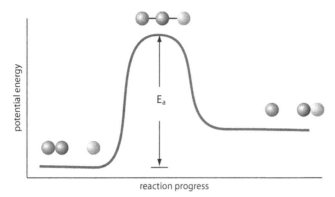

Figure 7.4 Reactive encounter.

Figure 7.5. All amino acids have two functional groups, carboxylic acid and amine; most have additional functional groups. The general formulas of all amino acids are much the same (proline is a little eccentric). They differ at one point called the R-group (Fig. 7.6). The R-group can be as simple as a hydrogen atom, in which case the amino acid is glycine, or it can be fairly complex. If the R-group is not hydrogen, then there are four different things attached to the central carbon: the R-group, a hydrogen atom, the carboxyl group (–COOH), and the amino group (–NH$_2$). A carbon with four different things is asymmetric, so each amino acid has a pair of enantiomers (except glycine). Living systems use only the L-enantiomer.

Protein Structure

An amine can react with a carboxylic acid to give an amide and water. Amino acids are both amines and carboxylic acids, so they can undergo this reaction linking two amino acids (Fig. 7.7). The product still has an unused amino group and an unused carboxyl group so additional amino acids can be attached at one or both ends. Long chains of amino acids, called proteins, or shorter chains called peptides can form. The sequence of amino acids is called the **primary structure** of the protein. Because the subunits of these chains can have different R-groups, a variety of properties is possible. There are R-groups that are nonpolar, R-groups that are polar, R-groups that are acids or bases, and R-groups that hydrogen bond. As a result, proteins can be made with polar regions, nonpolar regions, hydrogen-bonding regions, positively or negatively charged regions, and many others. The chains fold and twist to get the parts that attract one another close together and the parts that repel far apart. It is also possible to make covalent bonds from one part of the chain to another. The amino acid cysteine has the R group –CH$_2$SH. Two cysteines in different

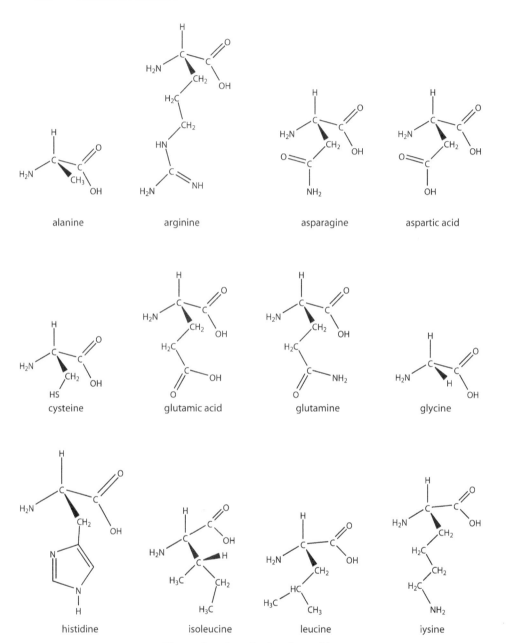

Figure 7.5 Standard amino acids.

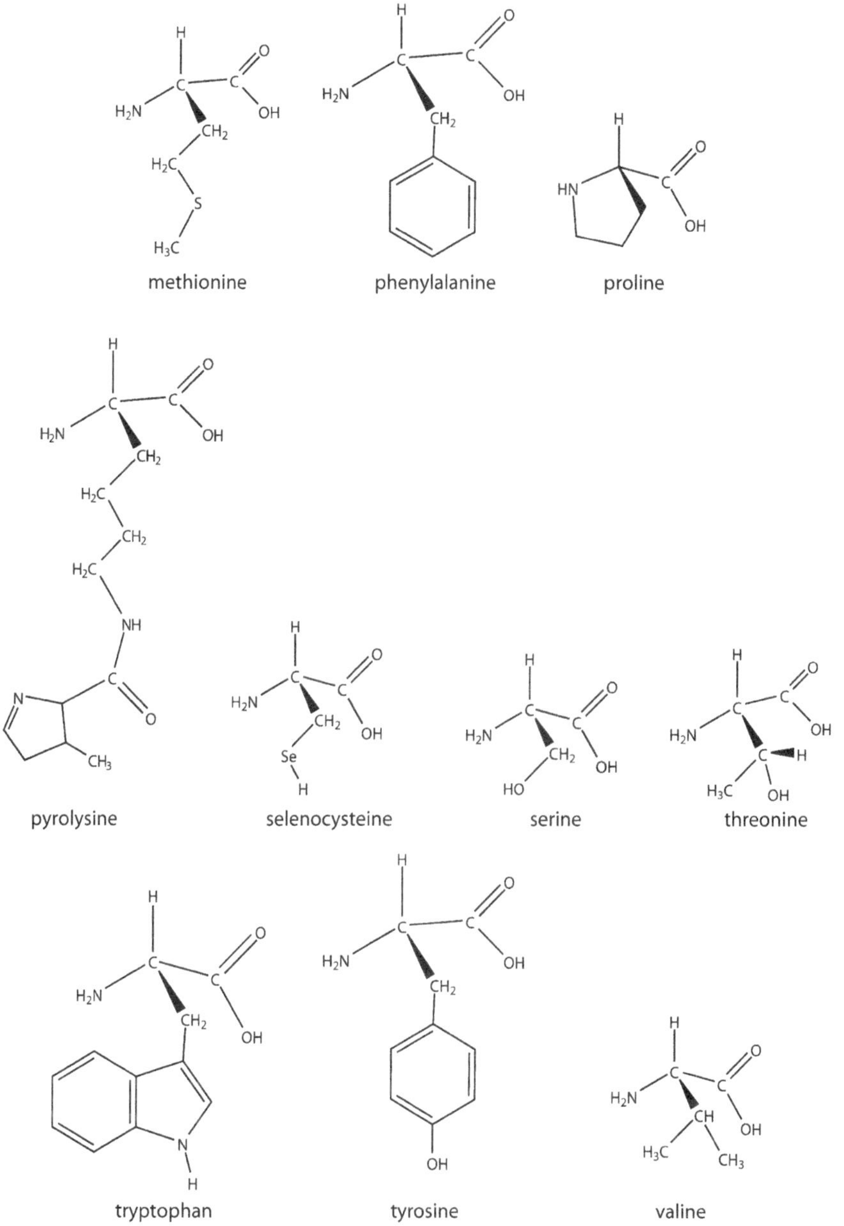

methionine phenylalanine proline

pyrolysine selenocysteine serine threonine

tryptophan tyrosine valine

Figure 7.5 (*Continued*)

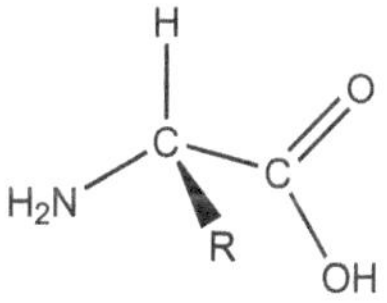

Figure 7.6 Amino acid.

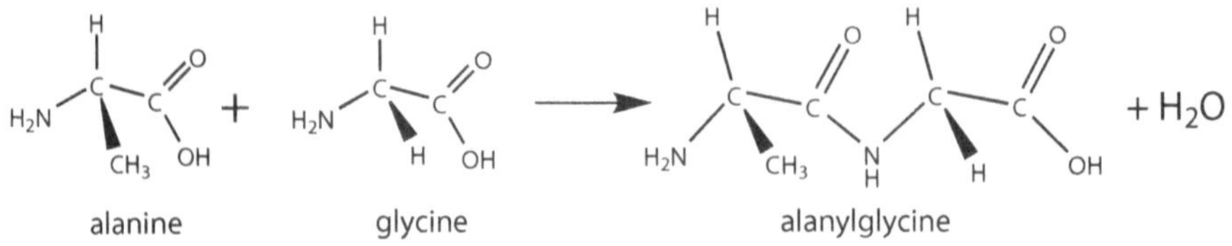

Figure 7.7 Amino acid condensation.

parts of the chain can bond together with a disulfide link, $-CH_2S-SCH_2-$. These interactions between regions on the protein chain can give the protein molecule a very specific shape. Pockets that bind only certain compounds or families of compounds can be made. Living systems take advantage of this versatility by producing proteins with special functions as diverse as providing outstanding toughness to spider silk to serving as highly active catalysts for chemical reactions.

Enzyme Function

Enzymes serve as biological catalysts for chemical reactions. The enzyme molecule has regions that bind reactant molecules in positions that allow them to react. Often part of the molecule is bound into a groove or depression in the surface of the enzyme molecule. Parts of the enzyme that interact with parts of the reactant molecules are called **binding sites**. Sometimes the binding of the reactant will cause changes in the shape of the enzyme to grip the reactant more tightly or to bend it to expose the target bonds. These adjustments are sometimes reminiscent of doors closing or tongs gripping. At the appropriate location on the enzyme, active groups approach the reactant bonds that are involved in the reaction. These active groups and their molecular environment make up the **active site** of the enzyme.

Figure 7.8 shows a simplified sequence for a catalytic reaction breaking a long molecule into two parts. The empty groove has binding sites and an active site. When the reactant enters the groove, the shape of the enzyme molecule changes, bringing the active site into position. This is called an **induced fit**. Cutting the molecule allows the groove to open back up, releasing the products. The model is not to be taken literally. There is no mechanical wedge slicing the reactant. Details for the process at the active site of the amylase enzyme are shown in Figure 7.9.

Enzyme binding sites can often bind a certain compound or class of compounds much more effectively than they can bind other compounds. As a result, the enzyme is active for certain reactants and nearly inactive for others, even though the potential reactants may be closely related. This is called **specificity**.

Living cells have thousands of enzymes, each of which has its own function. Even though the barley seeds in malt are dead, some of their enzymes are still active and continue to work during mashing.

MASHING PROCESS 141

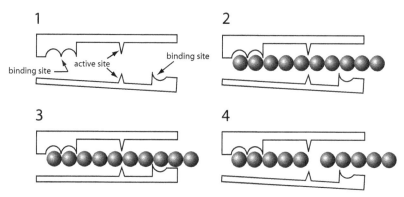

Figure 7.8 (1) Enzyme groove, (2) enzyme with reactant, (3) induced fit, and (4) products released.

AMYLASE MECHANISM

Enzyme catalysis occurs at a location called the **active site**. The R-groups of the amino acids are arranged in the active site to allow them to act on the bonds of the reactants. Amylase catalyzes the breaking of the glycosidic (glucose–glucose) bond, the breaking of the H–OH bond, and the addition of H and OH to the starch fragments. The sequence in Figure 7.9 shows a simplified mechanism for the hydrolysis process. The E's in the drawings represent the enzyme protein.

In addition to the active site, other parts of the enzyme give specificity by binding the correct reactant molecules and excluding others. Sometimes the binding strains certain bonds, making the reaction easier.

7.4 MASHING PROCESS

Amylase

Mashing involves adding hot water to the crushed grain so that the starch will gelatinize, and the amylase enzymes can do their work of liquefying and saccharifying (Fig. 7.10). The central technical issue of mashing is that beta-amylase steadily loses its activity in hot water. This happens to alpha-amylase as well, but more slowly at practical mashing temperatures. Activity loss occurs because the complex shapes of the enzyme molecules change in hot water. An enzyme that loses activity is said to become **denatured**. The hotter the water, the more rapidly the beta-amylase is denatured. The instability of beta-amylase is unfortunate, because beta-amylase most effectively nibbles away at the ends of the starch chains after the alpha-amylase has cut the chains up to make free ends. This leads to a dilemma for the brewer. If he or she mashes at a high temperature, like 158°F (70°C), the alpha-amylase will liquefy lots of starch, but the

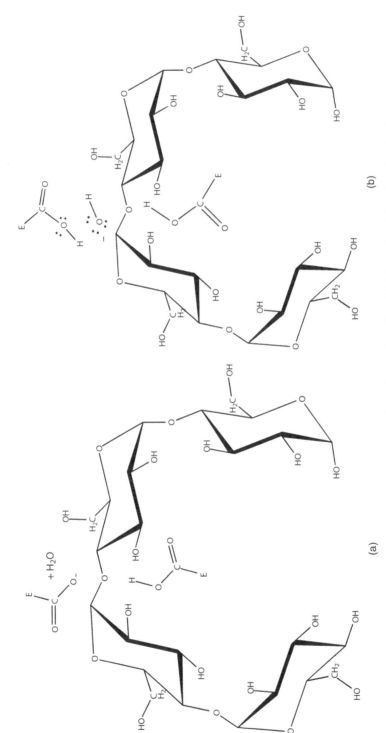

Figure 7.9 (a) R-groups in the active site include a carboxylic acid (glutamic acid) below the starch chain and a carboxylate (ionized aspartate) above the chain. (b) The carboxylate, acting as a base, pulls a hydrogen ion from water, leaving a hydroxide ion, OH⁻.

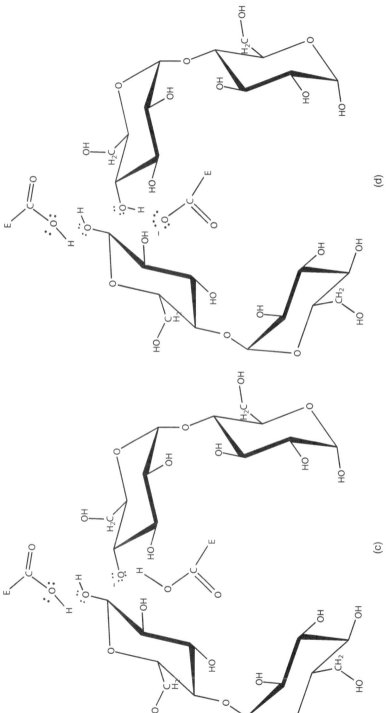

Figure 7.9 (*Continued*) (c) The hydroxide ion forms a bond to the anomeric carbon. This breaks the glycosidic bond to the next glucose unit. (d) A hydrogen ion from the carboxylic acid below transfers to the negatively charged oxygen atom, completing the hydrolysis.

Figure 7.10 Barley alpha-amylase. (*Source:* A. Kadziola et. al., *J. Mol. Biol.* **1994**, *239*: 104–121.) (See color insert.)

beta-amylase will be deactivated before it can saccharify all of it. The result will be high extract (most of the starch gets dissolved) but poor fermentability, because the glucose rings will still be tied together in chains that are too long for the yeast to use. If the brewer mashes at a lower temperature, like 140 °F (60 °C), the extract will be low but the fermentability will be high. Brewers usually seek a compromise between these extremes. One way to do this is to start mashing at a low temperature, then increase the temperature of the mash. Some breweries raise the temperature with the European practice of taking out some of the mash, heating it to boiling, and adding the boiling mash to the remaining mash. This straightforward process has the fancy name of **decoction**. Many breweries use a similar approach adapted to the use of unmalted grains, like rice or maize. The unmalted grain, called **adjunct**, is boiled in a separate cooker because of its high gelatinization temperature. The boiling hot adjunct is then added to the main barley mash. This raises the temperature of the mash and allows the malt enzymes to saccharify the starch from the adjunct.

Peptidase

In addition to starch conversion, mashing also converts protein from the grain. Protein is a mixed blessing in brewing. The protein can provide nutrients that the yeast needs, and protein is essential to maintain the foamy head that adds to the appeal of beer. On the other hand, certain types of proteins form particles in the beer that can make it hazy. Most beer drinkers like bright, clear beer, so brewers try to eliminate haze. Like starch, proteins are long chains. They can be hydrolyzed by enzymes called **peptidases**. The enzymes that cut the chains in the middle are **endopeptidases** (Greek: *endon*, inside). Those that nibble the ends are **exopeptidases** (Greek: *exo*, outside). Peptidases are abundant in green malt (malt before kilning), but most are destroyed by the kilning process. Any that remain are easily denatured at mashing temperature.

Sometimes the mash is started at low temperature, when denaturing is slower, to take advantage of any remaining peptidase activity. Most of the protein removal during mashing is not a result of enzyme action. Instead, the protein molecules become denatured, and the parts of the molecules that had been held together by intramolecular forces come apart, allowing the molecules to stick to one another (coagulate) and come out of solution.

Mash Tun

Mashing is carried out in a vessel called a **mash tun** or a mash conversion vessel (Fig. 7.11). In a commercial brewery the grist (crushed grain) is mixed with hot water in a pipe leading into the tun. Some mills are set up to add hot water before milling. Home brewers either add the grist to the water or add the water to the grist in the mash tun. About one to two quarts of water is used to mash one pound of grist (2–4 liters per kilogram). The ratio of the water volume to the grist mass is called the **mash thickness**. The mash tun has a stirrer (or workers with oars) for agitation. In some systems, the temperature

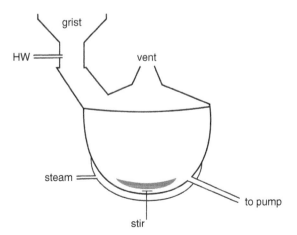

Figure 7.11 Mash tun.

can be raised with a steam or hot water jacket around the tun. Some homebrewers use a big pot for the mash tun, which allows them to provide heat from the stove top. Some traditional breweries have no means to raise the temperature of the mash other than by adding hot water (or hot mash). Many homebrewers have the same restriction because they use an insulated picnic cooler or a water cooler as a mash tun. Mashing usually takes about an hour, although a multistep decoction process can take as much as six hours.

7.5 DEXTRINS, LIGHT BEER, AND MALT LIQUOR

The amylase enzymes from malt do their best work on unbranched portions of starch chains. They do not hydrolyze the glycosidic links at or near branches. As a result, even after complete mashing, some of the carbohydrate is not fully converted into sugars that the yeast can ferment. The short, branched chains that remain are known as **dextrins**. They end up in the finished beer because the yeast does not convert them to carbon dioxide and alcohol. Dextrins are said to impart desirable properties to beer, such as mouth feel and body. They have no taste of their own, but they can break down to some extent under the influence of enzymes in the mouth to give sugars, which do affect the taste. If these dextrins could be made into fermentable sugars, the result would be more alcohol from the same malt (malt liquor), or the same alcohol from less malt (light beer). This can be accomplished to some extent by adopting special mashing procedures, but to get the very low levels of dextrins considered desirable in light beer, an enzyme called **amyloglucosidase** is used. This enzyme, typically derived from mold, is usually added to the beer during fermentation. Amyloglucosidase causes all glycosidic links, including those at branches, to be hydrolyzed. This converts dextrins to fermentable sugar.

Light beer is sometimes sold on the basis of having fewer calories than regular beer. The average non-light beer has about 153 food Calories in a 12 ounce (355 milliliter) serving. The same serving of the average light beer has about 103 Calories. By comparison, a regular-sized bagel has about 289 Calories (see Table 2.1). Substituting six regular beers with light beers would make up for eating a single bagel. Light beer should not be confused with low alcohol or alcohol-free beer. A typical regular American lager might be 5% alcohol by volume and the corresponding light version might be 4.2%. The alcohol itself provides 18.4 Calories for each percent alcohol in 12 ounces, or 92 Calories for 5% alcohol.

BIBLIOGRAPHY

Briggs, Dennis E.; Boulton, Chris A.; Brookes, Peter A.; and Stevens, Roger. *Brewing Science and Practice*. CRC Press, 2004, Chaps. 4–6. Detailed coverage of milling and mashing procedures and scientific background.

Kadziola, Anders; Abe, Jun-ichi; Birte, Svensson; and Haser, Richard. Crystal and Molecular Structure of Barley α-Amylase. *J. Mol. Biol.*, **1994**, *239*(1): 104–121. Source of the amylase structure in the Protein Data Bank.

Muller, Robert. The Effects of Mashing Temperature and Mash Thickness on Wort Carbohydrate Composition. *J. Inst. Brewing*, **1991**, *97*: 85–92.

Muller, Robert; and Canterranne, Evelyne. Activity of Amylolytic Enzymes in Thick Mash. *J. Am. Soc. Brewing Chem.*, **1994**, *52*(2): 56–61.

Narziss, L. The Influence of Mashing Procedures on the Activity and Effect of Some Enzymes. *MBAA Tech. Quarterly*, **1976**, *13*(1): 11–21.

QUESTIONS

7.1. What is the difference between a roller mill and a hammer mill?

7.2. What is the principal reaction accomplished by mashing?

7.3. What is hydrolysis?

7.4. What is a protein?

7.5. What is the role of enzymes? How do they help cells control chemical reactions?

7.6. How does the action of alpha-amylase differ from that of beta-amylase?

7.7. What is an adjunct?

7.8. What is the difference between extract and fermentability?

7.9. Explain light beer and malt liquor.

7.10. Draw the R-group on the amino acid leucine.

7.11. Identify two amino acids whose R-groups can form hydrogen bonds.

7.12. Explain the general mechanism of enzyme action. Be sure to mention the concepts of induced fit, binding site, and active site.

7.13. What is specificity? How does it come about?

*7.14. Draw the two dipeptides that could result from condensation of alanine with leucine.

7.15. Describe the events at the active site during amylase-catalyzed hydrolysis of a glycoside bond.

Lauter tun. Yuengling Beer Company, Pottsville, Pennsylvania.

CHAPTER 8

WORT SEPARATION AND BOILING

8.1 WORT SEPARATION

After mashing is complete and no starch worth recovering remains in the grist, the sugary liquid, called **wort** (pronounced "wert"), has to be separated from the bits and pieces of spent grain. This is called wort separation (Fig. 8.1).

Mash Tun Separation

In traditional British breweries, in small craft breweries, and in home breweries that don't buy already mashed extract, there is a strainer plate above the bottom of the mash tun. This plate, called the **false bottom**, has slots or holes like a colander to hold back the grain while the liquid goes through. The false bottom does not really filter the wort. The particles of spent grain that pile up on the false bottom serve as the actual filter. To set up the grain bed, wort is pumped from the bottom of the tun and recirculated until the wort runs clear, a process called **vorlauf**. Clear wort is then drawn through the grain bed, out the bottom of the mash tun, and into the kettle. If the grist had been ground very fine, the spaces between the particles would be very small, so flow would be too slow. Also, if the particles are nothing but mushy bits of grain without any of the hull, the flow could be blocked altogether. The objective of milling is to make grist that is fine enough so that the starch granules can gelatinize and liquefy well in a short time, but not so fine that the wort will not flow through it. Undamaged hulls are needed to lighten up the grain bed to get

---

The Chemistry of Beer: The Science in the Suds, First Edition. Roger Barth.
© 2013 John Wiley & Sons, Inc. Published 2013 by John Wiley & Sons, Inc.

Figure 8.1 Inside the lauter tun. Yuengling Beer Company, Pottsville, Pennsylvania. (See color insert.)

good flow. The amount of suction on the grain bed depends on the flow rate. If the wort is drawn off too quickly, the suction will compress the bed, crushing down the spaces, and cutting off the flow. After some or all of the liquid is run off, additional sugar is rinsed from the grain by adding hot water at the top. This process is called **sparging**.

Lauter Tun

The most common commercial system of wort separation is to pump the mash, grain and all, from the mash tun into a second vessel called the **lauter tun** (Fig. 8.2). The lauter tun has a false bottom like a British mash tun. The lauter tun is usually wide so the bed of grain is shallower than that in the mash tun, offering less resistance to flow. Inside the lauter tun is a set of knives that can be lowered into the grain bed to the desired depth. The knives move in a circle to cut into the grain bed and make filtration easier. A system of sprayers provides hot water to sparge the grain bed.

The mash is pumped into the lauter tun, recirculated for vorlauf, then the clear wort is run into the kettle or to a holding tank called the **underback**. The use of a lauter tun has the advantage that each of the two operations, mashing and wort separation, is carried out in a vessel that is optimized for that operation, rather than designed as a compromise. Also, capacity is increased because a new batch can be started as soon as the old one is sent to the lauter tun.

Mash Filter

A third system of wort separation is the **mash filter** (Fig. 8.3). The mash filter, in the form described here, is a relatively new method of wort separation

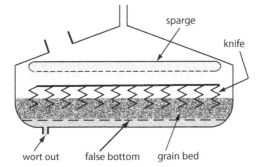

Figure 8.2 Lauter tun. (See color insert.)

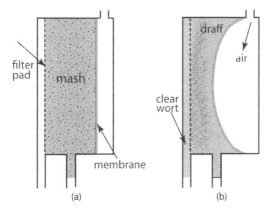

Figure 8.3 Mash filtration: (a) chamber with mash and (b) wort pressed out. (See color insert.)

pioneered by the Meura Company. The filter is a series of chambers, each of which is a box with a porous plastic filter pad covering one side. Behind the filter pad are channels to carry away the clear wort. The side opposite the filter pad has a balloon-like membrane that can be inflated with compressed air to drive the wort through the filter. The wort separation process starts with a batch of mash entering the filter chambers through pipes at the bottom. The mash pipes are closed off and the membranes are expanded with compressed air, driving the first wort through the filter pads into the channels. The membranes retract and water is fed in to sparge the grain. The membranes are expanded to drive the sparge water through the grain and the filters, and into the channels. The pressure provided by the inflated membranes can overcome the resistance even of finely milled grist. This is a big advantage of mash filtration because more sugar can be extracted from finely milled grist. In practice, mash filtration has yielded savings of 6.5% in malt consumption. Separation time is cut by 25–35% compared to lautering, increasing **brewhouse** capacity. Other advantages include lower water consumption and less water left in the draff. One disadvantage is the high cost of the filtration equipment. A second disadvantage is that mash filters need to be filled with a specific volume of mash. This detracts from the brewer's flexibility. Nonetheless, the method shows great promise.

8.2 BOILING

After the clear, sweet wort is separated from the grist, the wort is vigorously boiled, usually for about an hour, in a vessel called a **kettle** or a **copper**. The kettle needs some means of providing heat to the wort; commercial breweries often use steam pipes. Sometimes the actual heating and boiling take place in a device called a **calandria**. The calandria can be in the kettle, in which case the boiling itself propels the wort and causes it to circulate. Alternatively, the calandria can be outside the kettle, in which case the wort is pumped. A single external calandria can serve several kettles. In the photograph of the kettle at Victory Brewing in Figure 8.4, the external calandria is marked C. It is in the background at the lower right.

Boiling accomplishes five things.
- Resins from the hops change their molecular form and dissolve.
- Compounds that give bad flavors go off with the steam.
- Bacteria and wild yeasts in the wort are killed.
- Enzymes in the wort are deactivated by the heat.
- Proteins from the grain clump together so they can easily be removed.

Figure 8.4 Kettle at Victory Brewing Company, Downingtown, Pennsylvania.

The boiling process is set up to bring all parts of the boiling wort to the surface where it comes into contact with steam. This helps remove compounds, especially those containing sulfur, that can give bad tastes to the beer. During the boil, **volatile** compounds, that is, those that can vaporize, go off with the steam. The more contact the liquid makes with the steam, the more effectively undesired compounds are removed. One such compound is **dimethyl sulfide** $(CH_3)_2S$, which is formed from precursors in malt. It can give a cooked vegetable flavor to beer.

8.3 HOPS

Glands in the flowers of hop plants yield a yellow sticky powder called **lupulin**. Of the many components of lupulin, the most important is a family of compounds called **alpha acids**. Alpha acids are precursors to compounds called **iso-alpha acids** that give bitterness to beer. Iso-alpha acids also have some antibacterial properties that can help keep beer from spoiling. During boiling, the alpha acids are slowly converted to iso-alpha acids. This process, called **isomerization**, is shown in Figure 8.5.

The most abundant alpha acid is **humulone**, whose structure is shown in Figure 8.5. Small differences in the side chains (the groups hanging off the ring) result in compounds with different flavors. The iso-alpha acid corresponding to humulone is called **isohumulone** (Fig. 8.5). Alpha acids and iso-alpha acids have multiple functional groups; they are ketones, alcohols, and alkenes simultaneously. Under boiling conditions, no more than about 40% of the available alpha acids are converted to iso-alpha acids. In the end, only traces of alpha acids that have not been isomerized remain in the beer because unconverted acids are nearly insoluble at the cold temperature used for conditioning.

The target value for iso-alpha acids (IAAs) in beer is usually in the range of 10 to 60 milligrams per liter (0.0013 to 0.0080 ounces per gallon). Bitterness in beer is measured on a scale of **bitterness units** (BU), sometimes called international bitterness units (IBU), which are roughly equivalent to milligrams of isohumulone per liter. Because beer often contains other bitter compounds, the perceived bitterness can be higher than the BU value. The BU value is determined by a standard procedure involving absorption of ultraviolet light.

Hops can make contributions to the flavor apart from bitterness. Different varieties of hops have characteristic aromas that can become a part of the flavor of the beer. By definition, an aroma compound has to be easily vaporized so you can smell it. As a result, hop aroma compounds are generally lost on boiling. To get hop aroma, the brewer either adds hops after most of the boiling is finished (late hop addition), or after boiling and chilling (dry hopping). A less natural approach is the use of compounds dissolved from the hops with liquid carbon dioxide as discussed under Hop Extract below.

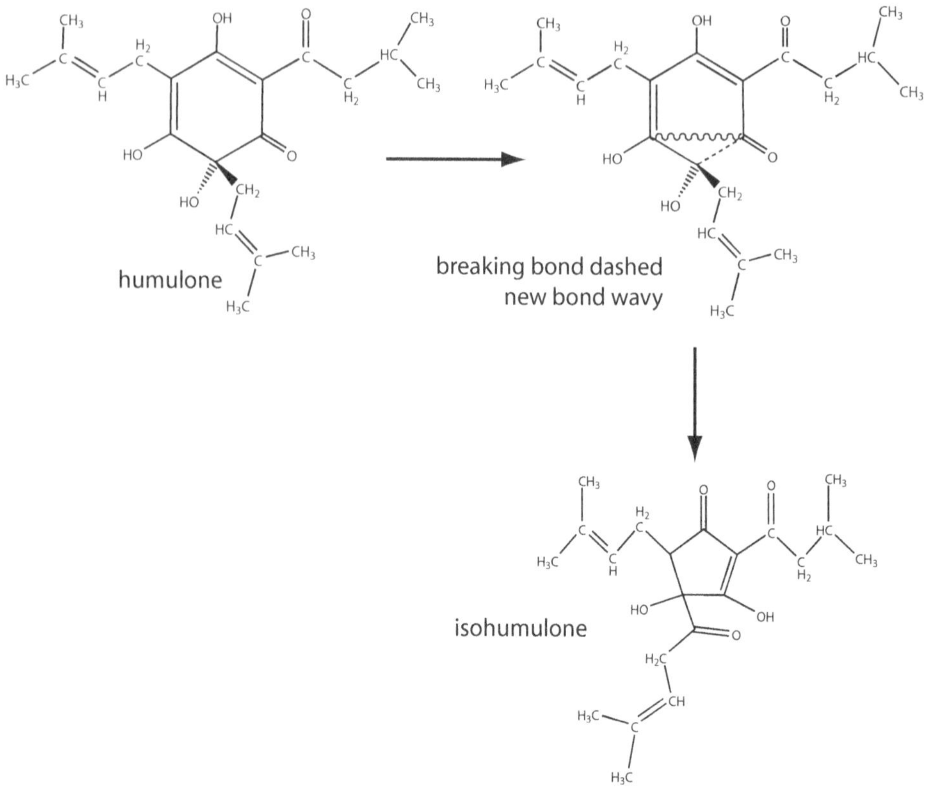

Figure 8.5 Humulone isomerization.

Iso-alpha acids resulting from the boiling of hops are sensitive to light. Exposure of hopped wort or finished beer to light whose wavelength is from 350 to 500 nanometers (near ultraviolet to green) results in a defect called lightstruck. Lightstruck beer smells like skunks. The mechanism for the production of lightstruck flavor is shown in Figure 11.13.

Hop Products

Hop cones are bulky and can be difficult to handle in automated equipment. In addition, they absorb valuable wort. A number of hop products are often used to make handling easier, to increase the fraction of the hop flavor compounds that actually end up in the beer (the utilization), and to protect the beer from being lightstruck.

Hop Pellets The simplest processing of hop cones involves grinding them to a powder, sometimes at low temperature, and pressing the powder into pellets. The powder may be sieved to concentrate the lupulin before pressing. Hops pellets take up less room than whole hop cones and they can be handled in automated equipment.

Isomerized Hop Pellets   A few percent magnesium oxide (MgO) is added to the hops powder before pressing it into pellets. The pellets are then heated without air. This causes a large fraction of the alpha acids to isomerize to iso-alpha acids. Isomerization allows a particular level of bitterness to be attained with less hops. The use of isomerized hop pellets allows the boiling time to be decreased, saving energy and lowering cost.

Hop Extract   Ground hops are treated with liquid carbon dioxide. The hop flavor compounds dissolve in this nonpolar solvent. After release of the carbon dioxide, the compounds form a thick oil called hop extract. Hop extract can be added directly to the kettle. This increases the fraction of the flavor compounds that actually end up in the beer, makes processing easier, and avoids addition to the kettle of insoluble plant material, which can be a source of off-flavors and haze.

Isomerized Kettle Extract   IKE and PIKE (a version neutralized with potassium hydroxide to enhance solubility) are liquid products made from hop extract in which the alpha acids have been pre-isomerized to iso-alpha acids. These have the advantages of isomerized hop pellets with the additional benefit of using a purified kettle addition.

Hydrogenated Extract   Hop extract or isomerized extract is treated to remove certain double bonds (Fig. 8.6). This inhibits the lightstruck reaction, making it possible to package the beer in clear bottles. Rho-isohumulone ("rho") results from treatment with sodium borohydride ($NaBH_4$). This adds two hydrogen atoms across the C=O double bond in the lower side chain to convert the ketone to an alcohol. Tetrahydro-isohumulone ("tetra") results from treatment with hydrogen and a catalyst. This adds four hydrogen atoms to convert the two C=C double bonds on the side chains to single bonds. Treatment with both sodium borohydride and hydrogen adds six hydrogen atoms that convert the ketone function and both double bonds. This product is called hexahydro-isohumulone. Tetra hydro- and hexahydro-isohumulone have the additional advantage of stabilizing beer foam.

The use of hydrogenated hop extract is characteristic of industrial brewing. Craft brewers usually use regular hop pellets or whole hop cones.

8.4   HOT BREAK

Another process in the boil is the removal of proteins. Barley protein can form small particles that make beer cloudy. In the wort, these proteins organize themselves so that the regions with many polar or hydrogen-bonding groups,

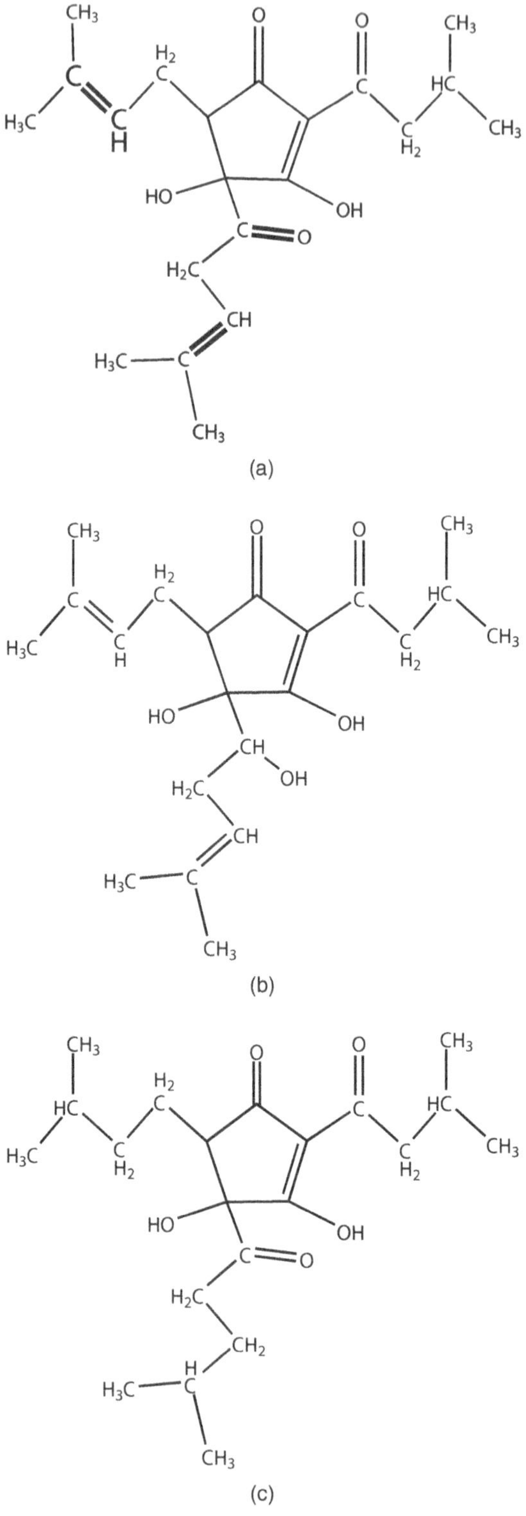

Figure 8.6 Hydrogenated hop extract: (a) isohumulone, hydrogenation sites marked; (b) rho-isohumulone; and (c) tetrahydro-isohumulone.

like –OH or –NH$_2$ (the hydrophilic regions), stick out and interact with the water. The nonpolar (hydrophobic) regions are on the inside and interact with one another. Under the influence of boiling, the protein molecules unwind so that the nonpolar regions of different protein molecules can come into contact. This causes the molecules to clump together or **coagulate** into large particles that can be removed. The coagulated protein formed during boiling is called the **hot break**. Proteins tend to migrate to the water's surface, which could be the surface of bubbles. This helps the protein molecules find one another and coagulate effectively.

Both the removal of off-flavor compounds and the coagulation of proteins occur faster when there is a large surface area of contact of the liquid wort with the steam. For this reason, vigorous boiling is needed to produce a large number of bubbles. In addition, many wort kettles are designed so that when the boiling wort comes out of the calandria, it splashes off of a deflection plate to scatter droplets throughout the vessel, which also increases wort contact with the steam.

After boiling, the hot break and remains of the hops are removed. Breweries that use whole hops remove them in a **hop-back**, a sort of strainer in which the hops serve as a filter. Hop aroma compounds can be introduced into the beer by adding unboiled hops to the hop-back. The remaining solids can be allowed to settle, or they can be collected in a **whirlpool** (Fig. 8.7). A whirlpool is a tank in which the hot wort is made to flow in a circle. This causes the solids to collect at the center where the motion is slowest. The **trub** (coagulated protein) and bits of hops collect in a pile at the bottom center of the tank (Fig. 8.8).

Figure 8.7 Whirlpool at Victory Brewing Company, Downingtown, Pennsylvania.

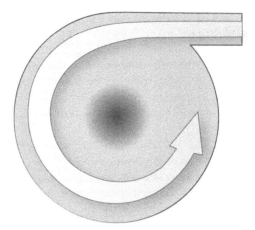

Figure 8.8 Whirlpool.

Figure 8.9 Countercurrent chiller.

8.5 CHILLING

After boiling, the wort must be chilled quickly from the boiling point at about 212 °F (100 °C) to the fermentation temperature of 45–72 °F (7–22 °C). Rapid cooling avoids the formation of off-flavor compounds. During chilling, additional protein, called **cold break**, coagulates. The **wort chiller** has a series of metal plates. Hot wort flows on one side of each plate and coolant flows on the other. As the wort passes through the chiller it cools down. As the coolant passes it warms up. It is most efficient to have the wort enter at one end of the chiller and the coolant at the other, which is called **countercurrent** flow (Fig. 8.9). This is because boiling hot wort will give up heat even to the warm exiting coolant. Cooler wort needs the entering cold coolant to accept its heat. The fundamental principle is that the flow of heat is proportional to the temperature difference. In a countercurrent chiller, the coldest coolant accepts heat from the coolest wort and the warmer coolant chills the hottest wort. Many breweries use hot water from the chiller as hot wash liquid or hot brewing liquor. This makes use of the energy in the hot wort, conserving resources and saving money.

BIBLIOGRAPHY

Briggs, Dennis E.; Boulton, Chris A.; Brookes, Peter A.; and Stevens, Roger. *Brewing Science and Practice*. CRC Press, 2004, Chaps. 7–10. All about hops and wort boiling.

De Keukeleire, Denis. Fundamentals of Beer and Hop Chemistry. *Química Nova*, **2000**, *23*(1): 108–112. Good summary of hop compounds, hop products, and the lightstruck reaction.

Hornsey, Ian S. *Brewing*. Hans Royal Society of Chemistry, 1999, Chap. 3. Hops, hop compounds, hop varieties, hop cultivation, and hop products available to brewers.

Krottenhaler, Martin; Back, Werner; and Zarnkow, Martin. Wort Production. In *Handbook of Brewing*, Hans Michael Eßlinger, Ed. WileyVCH, 2009. Plenty of detail about industrial brewery equipment and methods.

QUESTIONS

8.1. What is the purpose of sparging?

8.2. Describe two processes for wort separation.

8.3. What is lupulin?

8.4. What is hot break?

8.5. What happens to alpha acids during boiling?

8.6. What flavor do iso-alpha acids give to beer?

8.7. What is a calandria?

8.8. What five things does boiling accomplish?

8.9. What defect in beer is caused by dimethyl sulfide?

8.10. How are the various hop products made?

8.11. Which hop products would be most suitable for a beer that is to be sold in clear glass bottles?

8.12. Describe how a countercurrent chiller works.

8.13. For what is a whirlpool used?

*8.14. Draw the structure of hexahydro-isohumulone.

Fermenter. Victory Brewing Company, Downingtown, Pennsylvania.

CHAPTER 9

FERMENTATION

Fermentation turns sugar water into beer. During fermentation, a single-celled fungus called **yeast** feeds on sugar in the beer wort to get energy for its life processes. The overall chemical reaction for fermentation is

$$C_6H_{12}O_6 \text{ (glucose)} \rightarrow 2CH_3CH_2OH \text{ (ethanol)} + 2CO_2 + \text{energy}$$

The starting sugar may not actually be glucose, but the reactions are similar. The process occurs in a complex series of steps under the control of enzymes. For the yeast, the purpose is to extract energy in a form that the cells can use. The ethanol, although much prized by beer drinkers, is a waste product for the yeast.

Although the fermentation reaction does not consume oxygen, yeast cells require oxygen to make certain compounds that they need for growth and reproduction. Under certain conditions, yeast can engage in **respiration**, which does consume oxygen. The respiration products are carbon dioxide (CO_2) and water.

9.1 THE ANATOMY OF BREWING

To understand fermentation we need to step back and look at the general organization of a living cell, and at how materials and information get in and out of it. Any cell, including a yeast cell, is a bag whose contents are contained by a film called the cell membrane (also called the plasma membrane). The main structure of the membrane is made up of phospholipid molecules with

---

The Chemistry of Beer: The Science in the Suds, First Edition. Roger Barth.
© 2013 John Wiley & Sons, Inc. Published 2013 by John Wiley & Sons, Inc.

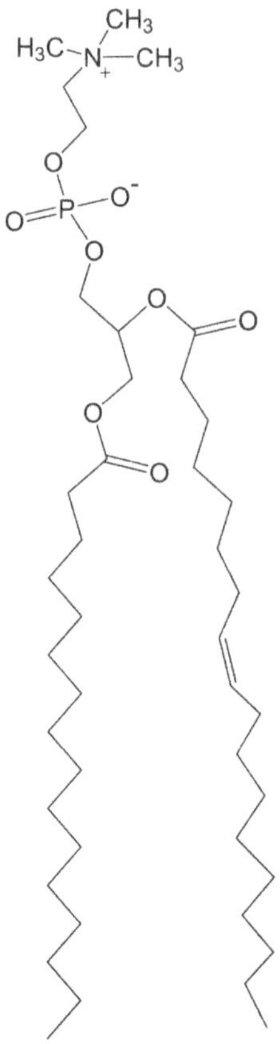

Figure 9.1 Membrane phospholipid.

highly polar, hydrophilic heads attached to long nonpolar hydrophobic tails, usually two tails per head. One member of this family of molecules is shown in Figure 9.1. In water, these molecules organize themselves into double sheets so that the tails point away from the water and the heads point toward the water. The membrane is a sandwich of nonpolar tails surrounded by polar heads, a structure called a **lipid bilayer** (Fig. 9.2). Polar molecules, including most things that dissolve in water, do not easily cross the membrane because of the nonpolar layer in the middle. If the hydrophobic tails in the lipid bilayer had no bends or kinks caused by double bonds, they would stick together because of dispersion forces, forming stiff, waxy membranes. Cell membranes are kept fluid by double bonds in the tail groups, which cause kinks in the chains, disrupting the dispersion forces (Fig. 9.3). In addition, compounds called **sterols**, which have an –OH group attached to a large nonpolar plate, get between the phospholipids, affecting the properties of the parts of the membrane that they occupy. Portions of the membrane that are rich in sterols tend to clump together forming **lipid rafts**. Lipid rafts serve as sites for protein structures attached to membranes. Figure 9.4 shows the most common sterol in yeast, ergosterol. Sterols are made from squalene ($C_{30}H_{50}$), an alkene with

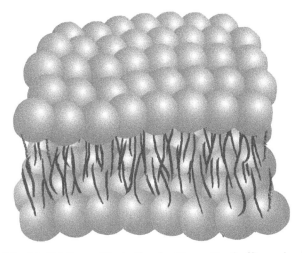

Figure 9.2 Lipid bilayer. Illustration by Marcy Barth. (See color insert.)

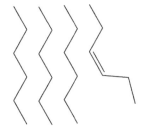

Figure 9.3 Tails with double bond.

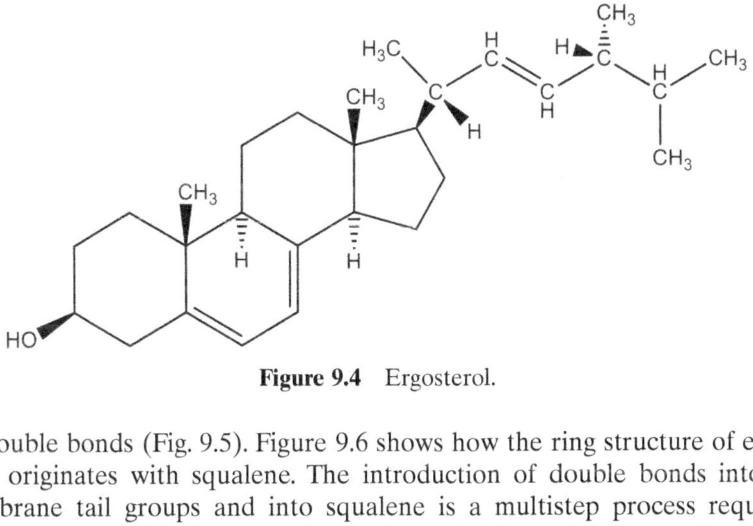

Figure 9.4 Ergosterol.

six double bonds (Fig. 9.5). Figure 9.6 shows how the ring structure of ergosterol originates with squalene. The introduction of double bonds into the membrane tail groups and into squalene is a multistep process requiring oxygen. The simplified net reaction is shown in Figure 9.7. Oxygen is needed at some point in the fermentation process to allow yeast to grow normally.

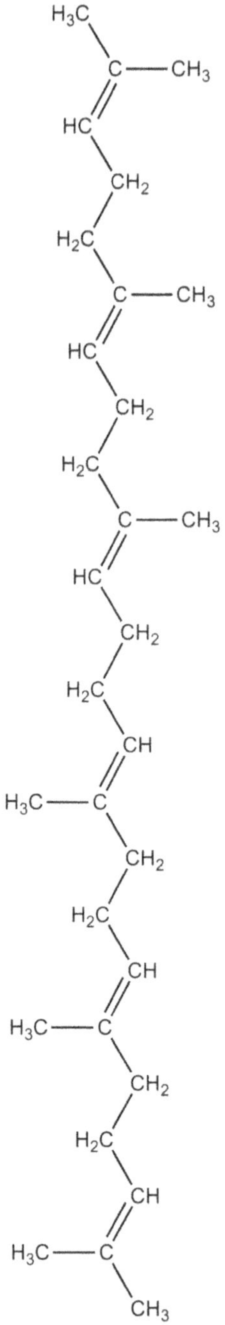

Figure 9.5 Squalene.

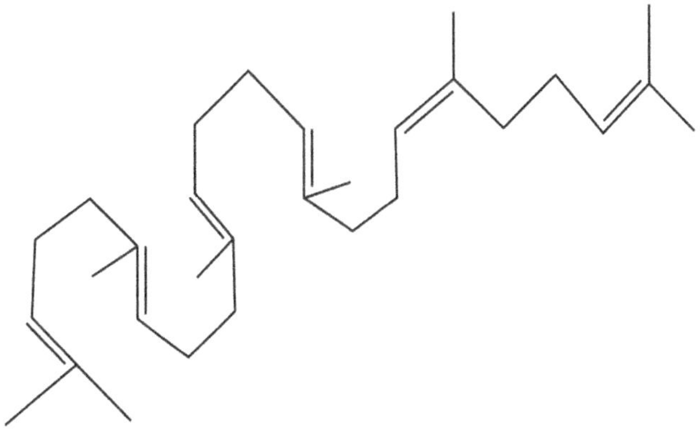

Figure 9.6 Squalene (folded).

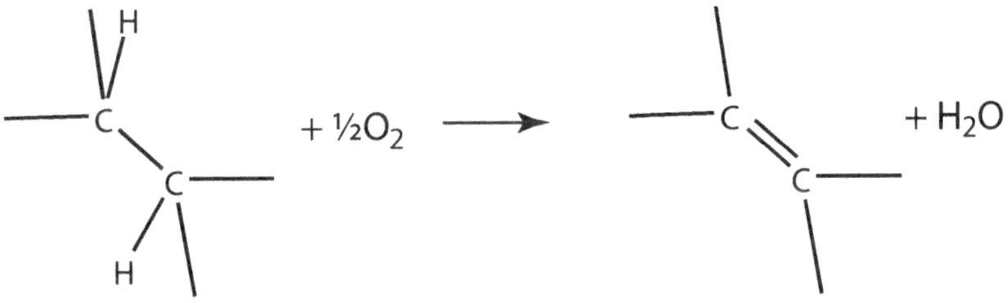

Figure 9.7 Making a double bond.

9.2 ENERGY AND BONDS

To understand fermentation, we need to look at **energy** and how it can be stored in and released from chemical **compounds**. The atoms that form a molecule are held together by **chemical bonds**. To break a bond takes work; energy must be supplied. This amount of energy is the **bond energy**. The stronger the bond, the larger the bond energy. Because energy is neither created nor destroyed, when a bond is formed the bond energy will flow out. The bond energy can be used to do work, like breaking other bonds, or it can be released as heat. If we run a reaction that breaks weak bonds and makes strong bonds, the energy we put in to break the weak bonds is small and the energy released by making strong bonds is large, so the process provides energy.

ATP

Adenosine triphosphate, ATP (Fig. 9.8), is the currency of energy in living cells. The way to think of this molecule is as a collection of modules (Fig. 9.9). The double ring at the upper right is adenine, which serves as an identifying tag to allow enzymes to recognize it. The single ring and the carbon to its left make

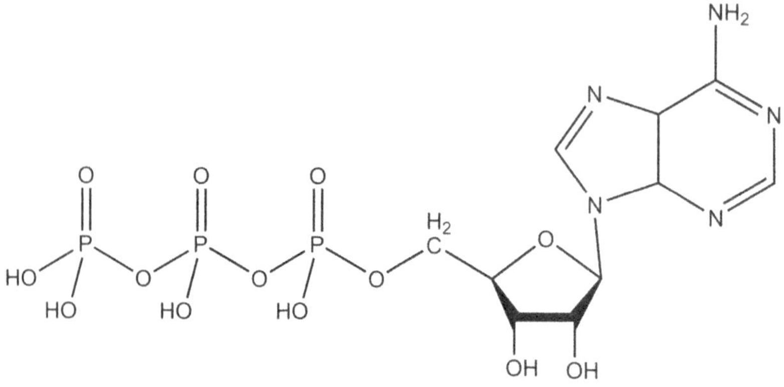

166 FERMENTATION

Figure 9.8 ATP.

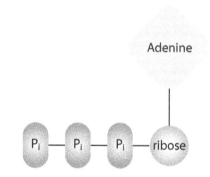

Figure 9.9 ATP modules. (See color insert.)

up a five-carbon sugar called ribose. The three connected phosphorus atoms are parts of phosphate groups. When one of the phosphate groups is hydrolyzed from ATP, giving adenosine diphosphate (ADP) and a molecule of phosphoric acid (H_3PO_4), energy is released (see box on "Energy from ATP"). This energy can be used to drive nonspontaneous processes. Energy is stored by the reverse reaction.

In a living cell, energy from the breakdown of food is temporarily stored in ATP molecules or wasted as heat. When the cell needs to use energy, ATP is hydrolyzed to ADP and phosphoric acid. When extra energy is available, energy from ATP is used to make sugars, starches, fats, or storage proteins that can be broken down to make more ATP when needed. The starch in barley seeds was produced by the plant to provide ATP for the embryo. Yeast cells use energy released by the fermentation of sugar to make ATP to run nonspontaneous life processes, like pumps to drive ions or molecules into or out of the cells.

ENERGY FROM ATP: A CLOSER LOOK

When ATP is hydrolyzed by water, an O–P bond in ATP and an O–H bond in water are broken. New O–P and O–H bonds are formed in the products. The broken and formed bonds are highlighted in Figure 9.10. In this diagram we explicitly show that, at the pH of yeast cells in beer (pH 4–5), the phosphate groups act as acids and give up some of their hydrogen ions. The energetics of ATP hydrolysis is subject to a good deal of confusion. As we can see, the bonds broken and the bonds made are the same. To understand why energy is made available by this reaction, we need to look more closely at the bonding in the hydrogen phosphate ion.

It is possible to draw the Lewis structure of the hydrogen phosphate ion, one of the products of ATP hydrolysis, in three ways that differ only by the placement of the electron pairs (Fig. 9.11). Because the forms are equivalent, there is no reason to favor one over the other. As a result, the actual structure is an average of these three forms, called **resonance** forms. Because of resonance, the negative charges are spread out over three oxygen atoms. In ATP the oxygen on the right on the hydrogen phosphate ion is bound to another phosphate, taking it out of the resonance picture and forcing the negative charges to be confined. Resonance allows the electrons in the unbound hydrogen phosphate that results from ATP hydrolysis to be delocalized so they can get farther apart, lowering their energy. This gives the products of ATP hydrolysis lower energy than the reactants. Resonance delocalization is the main source of the energy released when ATP is **hydrolyzed**.

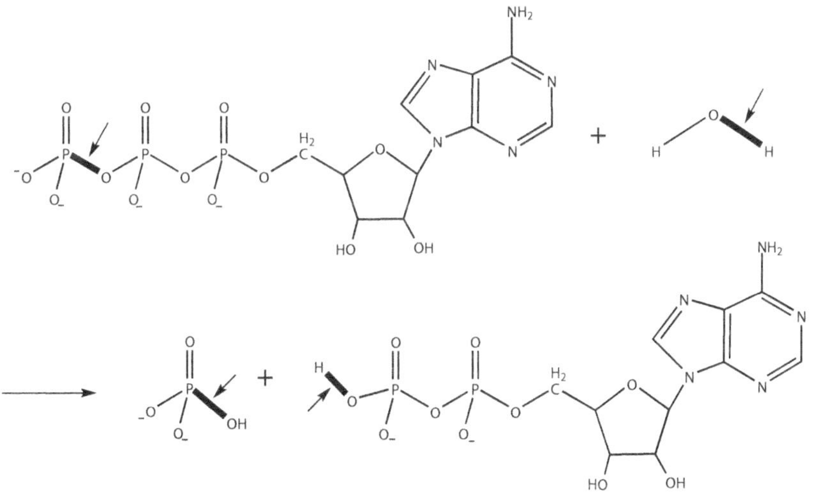

Figure 9.10 Hydrolysis of ATP.

(continued)

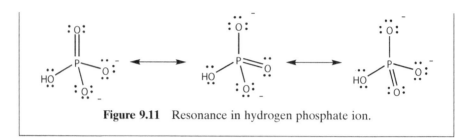

Figure 9.11 Resonance in hydrogen phosphate ion.

9.3 GLYCOLYSIS

The processing of glucose to pyruvic acid (see Fig. 9.16) is called **glycolysis**. The version of glycolysis used by yeast is called the **Embden–Meyerhof pathway**. We have simplified the scheme by combining some steps.

In the first step of glycolysis, shown in Figure 9.12, phosphate groups are attached at two places on a sugar molecule. These phosphates come from ATP, because energy is needed to accomplish these reactions.

In the next step, shown in Figure 9.13, the six-carbon sugar breaks into two three-carbon sugars, each with one of the phosphate groups. So far, two ATP molecules have been used up.

Now a second phosphate group is added to each of the two three-carbon sugars (Fig. 9.14), but this phosphate does not come from ATP. The second phosphate goes on to a C–H group rather than an O–H group. As a result, hydrogen is a product instead of water. The two hydrogen atoms from each phosphorylation are shown as [H].

The two phosphates on each of the two sugars are now attached to ADP to produce four molecules of ATP (Fig. 9.15) and two of water. What remains of the sugars is called **pyruvic acid**.

The net reaction so far can be represented symbolically:

$$C_6H_{12}O_6 + 2\,ATP + 2\,ADP + 2\,H_3PO_4 \rightarrow 2C_3H_4O_3 + 4\,ATP + 2\,H_2O + 4[H]$$

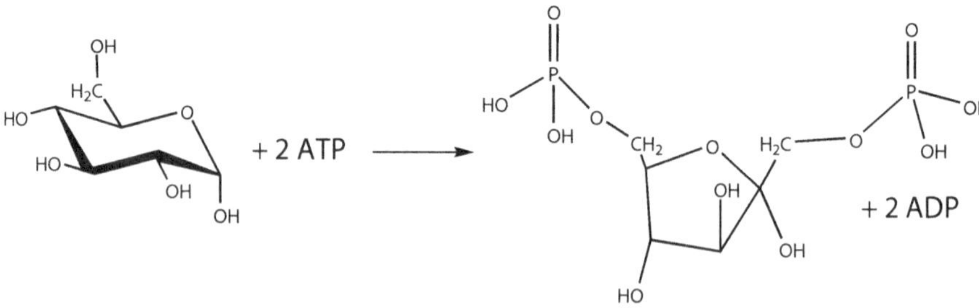

Figure 9.12 Phosphorylation.

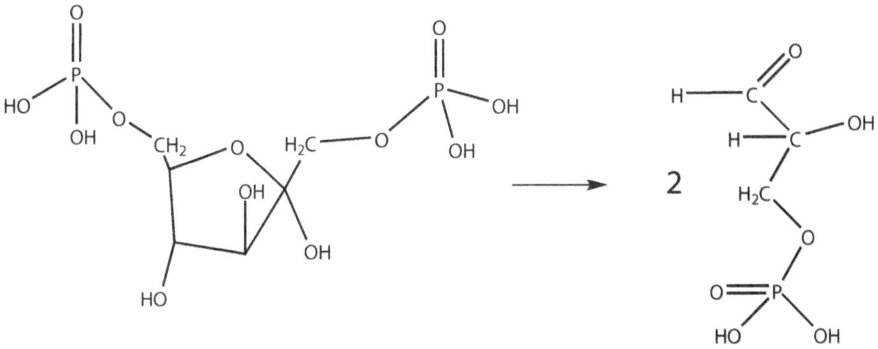

GLYCOLYSIS **169**

Figure 9.13 Sugar break up: fructose diphosphate → 3-phosphoglyceraldehyde.

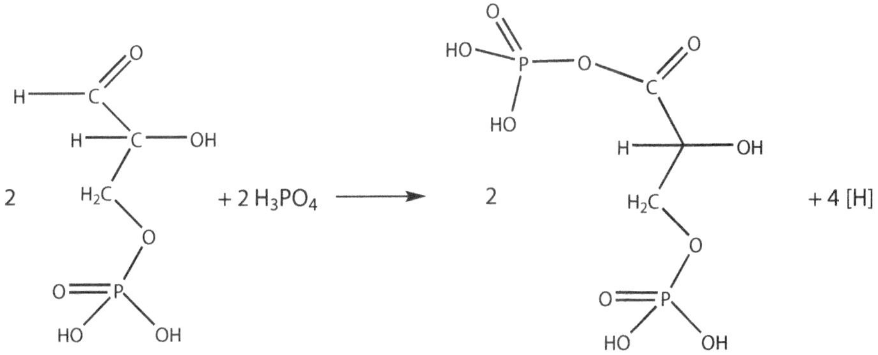

Figure 9.14 Phosphorylation with hydrogen transfer: 3-phosphoroglyceraldehyde → 1-3-diphosphoroglycerate.

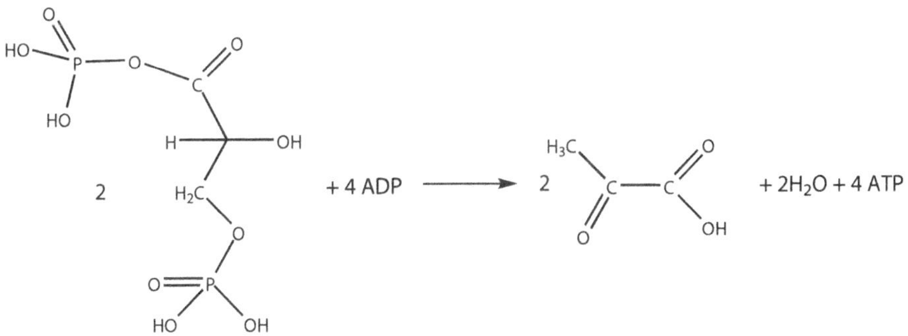

Figure 9.15 ATP production.

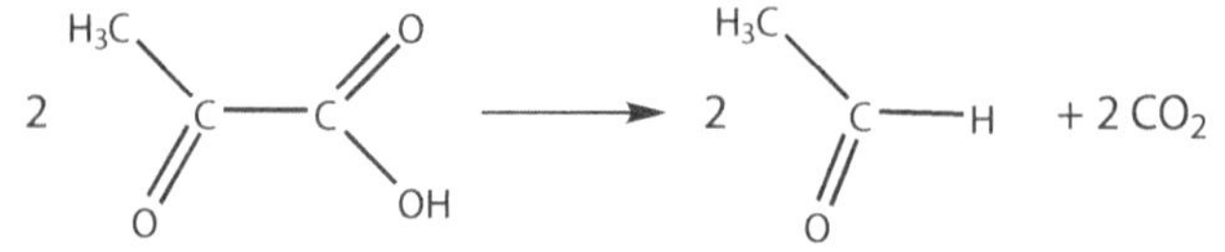

$$2 \underset{O}{\overset{H_3C}{\diagdown}}C-C\overset{O}{\diagup}_{OH} \longrightarrow 2 \underset{O}{\overset{H_3C}{\diagdown}}C-H + 2CO_2$$

Figure 9.16 Pyruvic acid decarbonylation: pyruvic acid → acetaldehyde.

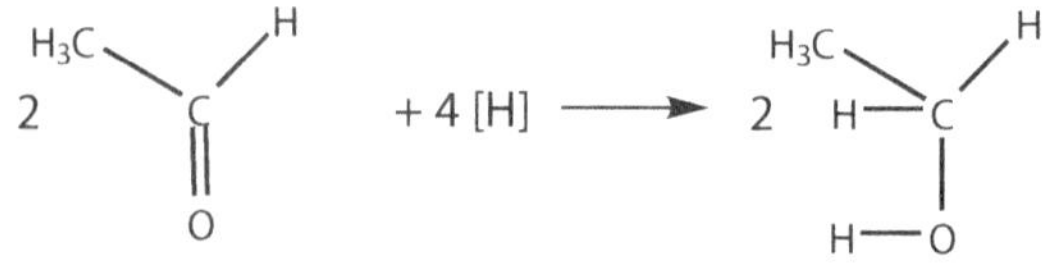

$$2 \underset{O}{\overset{H_3C}{\diagdown}}C\diagup^{H} + 4[H] \longrightarrow 2 \overset{H_3C}{\diagdown}H-\underset{H-O}{C}\diagup^{H}$$

Figure 9.17 Ethanol synthesis reaction: acetaldehyde → ethanol.

9.4 ETHANOL SYNTHESIS

The hydrogen atoms represented as [H] are parked temporarily on molecules called NAD^+ (nicotinamide adenine dinucleotide, don't ask). If oxygen were present, these hydrogen atoms could combine with the oxygen to give water and energy (ATP). There is no oxygen during fermentation so the following steps take place. One molecule of CO_2 is released from each pyruvic acid molecule, shortening the chain to two carbons and making a compound called acetaldehyde (Fig. 9.16).

The hydrogen atoms that were parked on NAD^+ are added to the C=O double bond of acetaldehyde to give ethanol (Fig. 9.17). This reaction takes up the hydrogen and regenerates NAD^+ so the process can continue.

9.5 AEROBIC AND ANAEROBIC REACTIONS

When oxygen is present and sugar levels are very low, yeast can feed on a number of compounds by a process called respiration. When the food source is glucose the respiration reaction is

$$C_6H_{12}O_6 \text{ (glucose)} + 6O_2 \rightarrow 6CO_2 + 6H_2O + \text{energy}$$

The same glycolysis pathway is followed, but the pyruvic acid then enters a pathway called the Krebs cycle that results in its complete conversion to CO_2 and water, with the production of several energy-rich molecules.

The fermentation reaction produces two molecules of ATP for each molecule of glucose, providing an energy equivalent of 340 kilojoules per kilogram of glucose. By contrast, respiration by yeast produces 28 molecules of ATP for each glucose molecule, which is equivalent to 4700 kilojoules worth of ATP per kilogram of glucose. So from the point of view of the yeast, the aerobic reaction provides 14 times as much usable energy as the anaerobic reaction. To the brewer the aerobic pathway is undesirable because it yields no alcohol. To the contrary, yeast can feed on alcohol by a respiration process, removing alcohol that had already been formed.

Yeast cells can live on the energy they get from fermentation, but some of their life processes go much better with oxygen. Oxygen is needed to make compounds with C=C bonds. Phospholipids with double bonds and sterols made from squalene, a hydrocarbon with six double bonds, are needed for membrane fluidity. The availability of these compounds makes the yeast healthier, more tolerant of high sugar and alcohol concentrations, and less prone to producing off-flavors. For this reason, oxygen is added to chilled wort before the yeast is added. Commercial brewers use pure oxygen to aerate the wort. Some home brewers pump filtered air into the wort, others just splash the wort about, which is better than nothing, but not really adequate for liquid yeast.

Crabtree Effect and Catabolite Repression

Yeast used in brewing has the unusual feature that even in the presence of air, alcoholic fermentation takes place and respiration is suppressed. The **Crabtree effect** (formerly called the *contre-Pasteur* effect) involves yeast that have been using oxygen for respiration when the concentration of sugar increases. The concentration of sugar exceeds the cells' capacity for respiration so fermentation takes up the excess. When the sugar concentration is high for a long time, as it is in beer brewing, the production of respiration enzymes and related compounds stops. This is called **catabolite repression**. Even in the presence of oxygen, the repressed cells will not engage in respiration.

Most organisms that can use oxygen will do so whenever it is available because respiration produces much more ATP for each sugar molecule than anaerobic fermentation. Brewing yeast, both *S. cerevisiae* (ale yeast) and *S. pastorianus* (lager yeast), forgo respiration and make alcohol when a sufficient concentration of sugar is present. Under these conditions, the organelles responsible for respiration are switched off. It has been shown that, in the presence of an adequate supply of sugar, the Crabtree effect and catabolite repression can confer an advantage to yeast in competition with other species. This is because the alcohol suppresses the growth of the competitors. When the sugar is nearly depleted the yeast resumes respiration and converts both the alcohol and any remaining sugar to carbon dioxide, water, and energy.

Under normal brewing conditions, respiration in yeast is repressed. If oxygen is available, it is used to make compounds that the yeast cells need for membrane fluidity and permeability (see Fig. 9.3, Fig. 9.4, Fig. 9.5, Fig. 9.6, and Fig. 9.7).

9.6 HIGHER ALCOHOLS

Alcohols with more than two carbon atoms, called higher alcohols or **fusel alcohols**, are produced during fermentation. Their concentration is higher when the fermentation temperature is higher. Fusel alcohols are implicated in hangovers produced by insufficiently aged distilled beverages. Their concentration in beer is rarely high enough to cause this problem, but they can affect beer flavor, usually contributing hot or solvent off-flavors. The real flavor issue is that alcohols, including fusel alcohols, can condense with carboxylic acids to make highly flavored esters that usually contribute fruit-like flavors. For example, isoamyl alcohol (systematic name 3-methyl-1-butanol) can be tasted at 70 milligrams per liter (70 ppm). When condensed with acetic acid, it produces isoamyl acetate, a banana-flavored ester that can be tasted at 1.6 milligrams per liter.

Yeast cells make fusel alcohols when they break down amino acids to get nitrogen, which they use to make new amino acids. The process is shown in Figure 9.18 with leucine as the amino acid, giving isoamyl alcohol. The portion of the amino acid that ends up in the higher alcohol is highlighted. The first step is called transamination. The amino group ($-NH_2$) on one molecule, the amino acid leucine in this case, trades places with the carbonyl oxygen ($=O$) on another molecule, 2-oxoglutaric acid (also called alpha-ketoglutaric acid). The products are 2-oxo-4-methylpentanoic acid and the amino acid glutamic acid. Notice that there is some similarity between 2-oxo-4-methylpentanoic acid and pyruvic acid (Fig. 9.16), a product of glycolysis. In both molecules the first carbon is part of a carboxylic acid group (COOH), the second carbon is part of a carbonyl group ($C=O$), and the third carbon is part of an alkyl group. As with pyruvic acid, the next step, shown in Figure 9.19 is the removal of CO_2 leaving an aldehyde. The aldehyde then picks up hydrogen from NADH to give an alcohol, as shown in Figure 9.20.

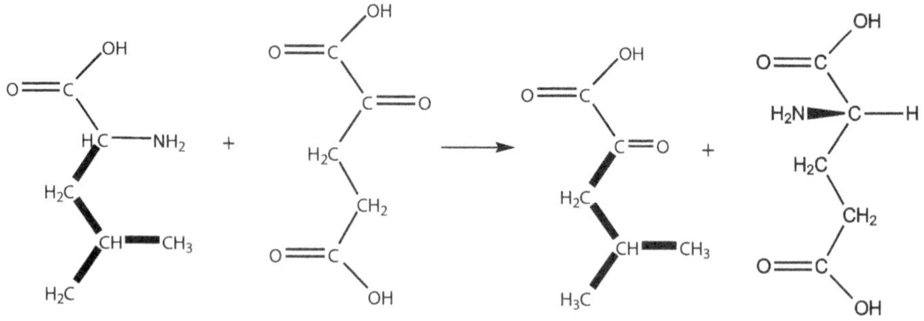

Figure 9.18 Transamination: leucine + 2-oxoglutaric acid → 2-oxo-4-methylpentanoic acid + glutamic acid.

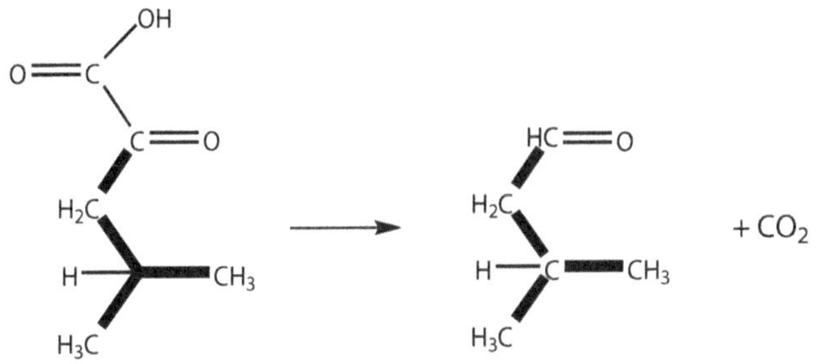

Figure 9.19 CO_2 removal from 2-oxo-4-methylpentanic acid: 2-oxo-4-methylpentanoic acid → 3-methylbutanal.

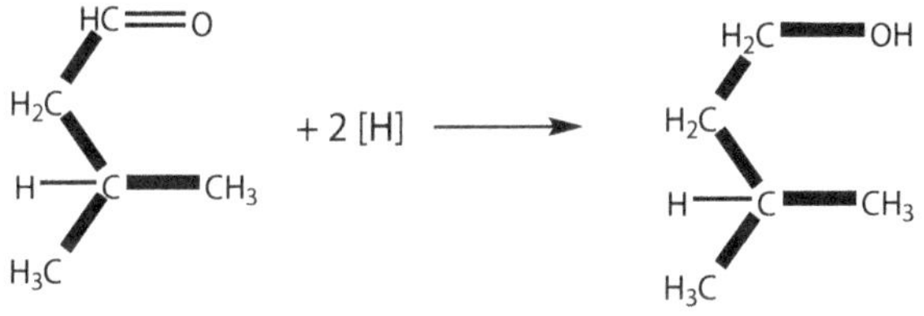

Figure 9.20 Aldehyde hydrogenation: 3-methylbutanal → isoamyl alcohol.

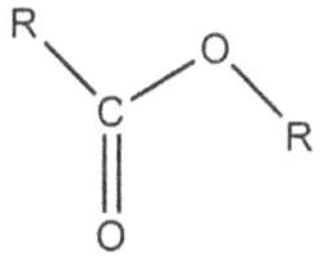

Figure 9.21 Ester.

9.7 ESTERS

Yeast cells produce a number of minor products that can have a big effect on the beer's flavor. The distribution of these products depends on the particular strain of yeast and on the fermentation conditions. One important family of compounds is the **esters** (Fig. 9.21). Esters usually give flower or fruit flavors to beer (and to fruits and flowers). This is considered good in most ale styles, but not in typical lager beers. An ester can be made by a condensation reaction from a carboxylic acid and an alcohol. In beer the acid part of the most common esters is acetic acid. The alcohol part can come from ethyl alcohol or fusel alcohols usually made in trace amounts during fermentation. The most prevalent ester is ethyl acetate, for which the acid is acetic acid (CH_3COOH) and the alcohol is ethyl alcohol (CH_3CH_2OH). Ethyl acetate has a flowery

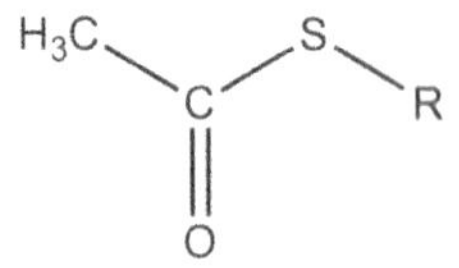

$$\text{H}_3\text{C} - \underset{\underset{\text{O}}{\parallel}}{\text{C}} - \text{S} - \text{R}$$

Figure 9.22 Acetyl coenzyme A.

aroma at low concentrations that becomes solvent-like at higher concentrations. The acetate esters (esters based on acetic acid) in beer are not made directly from acetic acid. Instead, they are made by switching an alcohol with one that is part of an acetate ester that is already present, called **acetyl coenzyme A**. The acid part of acetyl coenzyme A is acetic acid; the alcohol is actually a **thiol** because it has an –SH group instead of an –OH group. The business end of acetyl coenzyme A is shown in Figure 9.22, where R represents a complicated structure whose formula is $C_{21}H_{35}N_7O_{16}P_3$.

The types of esters made and their concentrations are influenced by the particular strain of yeast and the fermentation temperature. These two factors are related. Different strains of yeast perform best at different temperatures. Lager beer, fermented at low temperature with bottom-fermenting yeast, usually has a very low ester concentration, giving it a characteristic clean flavor. Ales, which are fermented at higher temperature with top-fermenting yeast, often have the fruity or flowery notes characteristic of esters. This is a clear example of the diversity of beer styles. Flavors that are appropriate and desirable in one style of beer may be considered defects in another style.

BIBLIOGRAPHY

Alvarez, Francisco J.; Douglas, Lois M.; and Konopka, James B. Sterol-Rich Plasma Membrane Domains in Fungi. *Eukaryotic Cell*, **2007**, *6*(5):755–763.

Goddard, Matthew R. Quantifying the complexities of *Saccharomyces cerevisiae*'s Ecosystem Engineering via Fermentation. *Ecology*, **2008**, *89*(8): 2077–2082. Yeast gets an advantage by making alcohol.

Hough, J. S. *The Biotechnology of Brewing and Malting*. Cambridge University Press, 1985.

Lehninger, Albert H. *Bioenergetics*. Benjamin-Cummings, 1971. How ATP works.

Van Laere, S. D. M.; Verstrepen, K. J.; Thevelein, J. M.; Vandijck, P.; and Delvaux, F. R. Formation of Higher Alcohols and their Acetate Esters. *Cerevisia*, **2008**, *29*(2): 65–81.

Walker, Graeme, M. *Yeast Physiology and Biotechnology*. Wiley, 1998. How things work in a yeast cell.

White, Chris; and Zainasheff, Jamil. *Yeast*. Brewers Publications, 2010. Covers the fermentation process, and the characteristics of yeast and other microbes relevant to brewing.

QUESTIONS

9.1. Write the overall reaction for fermentation. Give the names of each reactant and product.

9.2. Explain the difference between fermentation and respiration.

9.3. What is the function of ATP in cells?

9.4. Draw the structures of pyruvic acid, acetaldehyde, and ethanol.

9.5. What benefit do yeast cells get from fermentation?

9.6. What is the biological purpose of alcohol synthesis?

9.7. Describe the Crabtree effect.

9.8. What is a lipid?

9.9. How is beer affected by the presence of esters?

9.10. How would you adjust the fermentation conditions to give less fruity character in a beer?

9.11. Why does yeast need oxygen when it is first pitched?

***9.12.** Draw the higher alcohol that would result from valine.

Valine

***9.13.** Draw the ester that would result from the condensation of the alcohol from Question 9.12 with acetic acid.

Acetic acid

Fermenter controller. Victory Brewing Company, Dowingtown, Pennsylvania.

CHAPTER 10

TESTS AND MEASUREMENTS

Beer production at any level requires chemical, physical, and sometimes biological testing at various stages to assure quality and uniformity of the product and to avoid waste. Standardized apparatus and procedures go a long way toward maintaining product quality. Nonetheless, beer is ultimately judged by the extent to which it appeals to consumers. For this reason, beer is always tasted. Because the evaluation of flavor is individual and subjective, special procedures are needed.

10.1 CARBOHYDRATE CONTENT

Brewing is a process of converting starch into dissolved carbohydrates and then converting the carbohydrates into alcohol and carbon dioxide. The carbohydrate content of wort can be an indication of the efficiency of mashing. Carbohydrate content is used to monitor the progress of fermentation and to tell when it is complete.

Specific Gravity

The standard way to test for carbohydrate content is by **density** or **specific gravity**. Density is the mass of a material that occupies a particular volume. The density of water at 72 °F (22 °C) is 0.998 grams per cubic centimeter or 8.33 pounds per U.S. gallon. When sugar is dissolved in the water, the density increases by a small but measurable amount. In brewing, it is customary to use the specific gravity instead of the density. Specific gravity is the mass of a certain

---

The Chemistry of Beer: The Science in the Suds, First Edition. Roger Barth.
© 2013 John Wiley & Sons, Inc. Published 2013 by John Wiley & Sons, Inc.

volume of a substance divided by the mass of that same volume of water. The specific gravity of pure water is, by definition, exactly one. Specific gravity is close to density in grams per milliliter. In brewing, the specific gravity is often reported in **points**, which are the thousandths of the specific gravity in excess of one. Formally: points = (SG − 1) × 1000; a specific gravity of 1.044 would be 44 points. The specific gravity of water increases by about 0.004 (4 points) for each 1% of dissolved carbohydrate (see Table 10.2 for more detail). The specific gravity of wort before fermentation is called the **original gravity** or **OG**; the specific gravity after fermentation is called the **final gravity** or **FG**. These numbers can be used to estimate the alcohol content of the beer (see Table 10.3). Specific gravity is not selective for carbohydrates; anything dissolved in water affects the density of the solution. Fortunately, before fermentation beer wort has only small (but important) traces of anything other than water and carbohydrate. After fermentation there is enough alcohol to have a significant effect on the specific gravity.

Specific gravity permeates the jargon of beer styles. A beer made from wort with a high initial gravity is sometimes called a high gravity, or heavy beer. Original gravity is a key characteristic defining a beer style. Three popular British ale styles are distinguished by their original gravity ranges: Ordinary Bitter (OG: 1.032–1.040), Special Bitter (OG: 1.040–1.048), and Extra Special Bitter (OG: 1.048–1.060).

The usual way to measure specific gravity is with a **hydrometer**, a device that floats in the liquid (Fig. 10.1). Floating is governed by **Archimedes principle**, which states that a submerged object will experience a force (buoyancy) equal to the weight of the liquid that it displaces (pushes aside). If the object floats, the buoyancy is equal to the weight of the object. For example, if we have a cylinder weighing 10 grams with a cross-sectional area of 1 square centimeter, and we float it in a liquid whose density is 1 gram per cubic centimeter, it will have to sink until it displaces 10 grams of solution. So it would have to sink in to a depth of 10 centimeters (1 square cm × 10 cm = 10 cubic cm). If the liquid had a density of 1.05 grams per cubic centimeter, the cylinder would still have to displace 10 grams of it, but now the volume would be 10 grams ÷ (1.05 grams/cubic cm) = 9.52 cubic centimeters, so the cylinder sinks in by only 9.52 centimeters. If we put markings on the cylinder, we could read the density, or the specific gravity. Practical hydrometers accentuate this effect by having a big bulb weighted at the bottom that goes into the liquid and a narrow stem that sticks up. A small change in displacement results in a large change in height, making the hydrometer easier to read. Hydrometers give accurate results for the carbohydrate content of wort, but the results are not exactly correct for fermented beer. This is because the alcohol in fermented beer has a significant effect on the specific gravity. Up to 9% alcohol, the specific gravity is lowered by about 1.35 points for each percent alcohol by volume at 20 °C. To approximate the carbohydrate content we can correct the hydrometer reading by adding back in the 1.35 points for each percent alcohol by volume. The carbohydrate content can then be read from Table 10.2.

Figure 10.1 Hydrometer. (See color insert.)

HYDROMETERS: A CLOSER LOOK

Reading a Hydrometer

The first concern in using a hydrometer is to read the correct number from the scale. Water climbs up onto glass surfaces making a shape called a **meniscus** (Greek: *meniskos*, crescent moon). The correct reading is at the bottom of the meniscus, away from where it touches the glass. Don't forget that the numbers increase going down. In Figure 10.2, the dashed line shows the level at which the hydrometer is read. The hydrometer is showing a specific gravity of 1.037, or 37 points.

(continued)

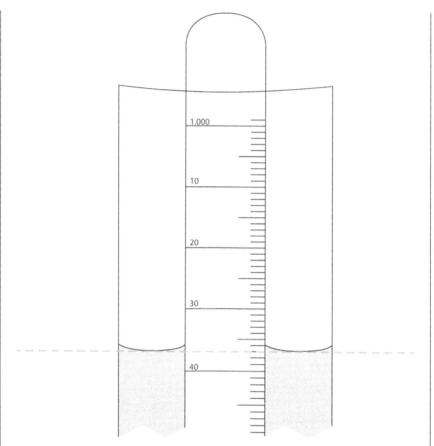

Figure 10.2 Reading a hydrometer. (See color insert.)

Temperature Correction

Suppose we made a bottle whose capacity is exactly one kilogram of water at 15°C (59°F). The mass (in kg) of any sample that exactly fills the bottle would be equal to the specific gravity of that sample *at 15°C*. But what if the temperature was not 15°C? Suppose it was 45°C. The bottle would not hold a whole kilogram of water at 45°C, because water expands as temperature increases. The mass of the sample would be less than the true specific gravity because we are using too small a bottle. For the same reason, if a hydrometer is used for a sample that is not at the temperature for which the hydrometer was designed, a correction must be applied. It is best to take the reading as close as possible to the temperature for which the hydrometer was designed. If that is not convenient, Table 10.1 gives an approximate correction (in points) that must be added to the reading for 15°C or 59°F hydrometers.

TABLE 10.1 Hydrometer Correction (Points)

°C	Correction	°C	Correction	°F	Correction	°F	Correction
5	−1	40	+7	40	−1	110	+8
10	−1	45	+9	50	−1	120	+10
15	+0	50	+11	60	+0	130	+13
20	+1	55	+13	70	+1	140	+16
25	+2	60	+16	80	+3	150	+19
30	+3	65	+19	90	+4	160	+22
35	+5	70	+21	100	+6	170	+26

THE HYDROMETER SCALE

No single number can give a full picture of beer wort composition. Nonetheless, given that all carbohydrates have about the same effect on the specific gravity, and that carbohydrates are the dominant nonwater component, the specific gravity provides a reasonable estimate of the dissolved solids in beer wort and other sugar–water solutions. The hydrometer scale is the mass percent of sucrose that has the indicated specific gravity. In Table 10.2, the SG column is in points and the data apply to 20 °C (68 °F). Suppose a 15 °C hydrometer is read at 20 °C giving a reading 1.041 = 41 points. The hydrometer correction table (Table 10.1) gives a correction of +1 point giving a corrected SG of 41 + 1 = 42 points, which corresponds to 10.48% extract.

TABLE 10.2 Hydrometer Scale

SG	Percent	SG	Percent	SG	Percent	SG	Percent
1	0.25	21	5.33	41	10.24	61	14.98
2	0.51	22	5.58	42	10.48	62	15.21
3	0.77	23	5.83	43	10.72	63	15.45
4	1.03	24	6.08	44	10.96	64	15.68
5	1.28	25	6.33	45	11.20	65	15.91
6	1.54	26	6.58	46	11.44	66	16.14
7	1.80	27	6.82	47	11.68	67	16.37
8	2.05	28	7.07	48	11.92	68	16.60
9	2.31	29	7.32	49	12.16	69	16.83
10	2.56	30	7.56	50	12.40	70	17.06
11	2.81	31	7.81	51	12.63	71	17.29
12	3.07	32	8.05	52	12.87	72	17.51
13	3.32	33	8.30	53	13.11	73	17.74
14	3.57	34	8.54	54	13.34	74	17.97
15	3.82	35	8.79	55	13.58	75	18.20
16	4.07	36	9.03	56	13.81	76	18.43
17	4.33	37	9.27	57	14.05	77	18.65
18	4.58	38	9.52	58	14.28	78	18.88
19	4.83	39	9.76	59	14.52	79	19.10
20	5.08	40	10.00	60	14.75	80	19.33

Figure 10.3 Refractometer.

Refractive Index

Another way to measure carbohydrate content is with **refractive index**. Refractive index is the ratio of the speed of light in a vacuum to that in a particular material. The more sugar is in a solution, the slower light travels in it, so the higher is the refractive index. The refractive index is measured by the amount the light bends as it passes from the sample to a plate of glass. The instrument used to measure refractive index is a **refractometer** (Fig. 10.3). The advantage of a refractometer is that it only takes a few drops of wort to make a measurement. As with specific gravity, the refractive index is not selective for carbohydrates. Alcohol has a large effect on the refractive index, so the method can only be used on unfermented wort.

CARBOHYDRATE CALCULATIONS

Brewing has a special jargon for the carbohydrate concentration and its changes during fermentation. The mass percent of solids in beer wort before fermentation is called the **original extract** (OE). The mass percent solids in decarbonated beer after fermentation is called the **real extract** (RE). The difference between the OE and the RE is called the **attenuation**. The attenuation is usually expressed as a percentage of the OE. If the wort has 12% solids before fermentation and 4% after, 8 g/100 g is attenuated. The degree of attenuation is 8% ÷ 12% = 0.67 or 67%. The attenuation is close to but not exactly the same as the amount converted by fermentation, which we will call the **degree of fermentation**. For each gram of solid converted, 0.4839 g of alcohol is produced and 0.5161 g of material bubbles out as carbon dioxide or drops out as yeast. If we had 100 grams of 12% OE wort and the degree of fermentation was 8 g/100 g, 8 g × 0.5161 = 4.1 g would be lost, leaving 95.9 g of beer. The RE would be 4 g ÷ 95.9 g × 100% = 4.17% and the attenuation would be 12 g/100 g − 4.17 g/100 g = 7.83 g/100 g, which comes to 65% of the OE of 12 g/100 g.

10.2 TEMPERATURE

Temperature is a measure of how much energy molecules have, that is, how fast they are moving. Chemical reactions usually go faster as the temperature increases. Enzyme molecules, which are needed to speed up and direct the brewing reactions, can shake themselves apart if the temperature is too high. There are temperature ranges in which each enzyme works best. Temperature measurement and control is essential, especially during mashing. When a temperature measuring device is used to give a numerical value for the temperature, it is called a thermometer; when it is used as input to control a heater or cooler, it is called a probe. We will lump them all together as thermometers. There are several types of thermometers available to brewers.

Liquid-in-Glass Thermometer

The liquid-in-glass thermometer depends on the expansion of the column of liquid, like mercury or alcohol, in a thin channel in a glass tube (Fig. 10.4). Liquid-in-glass thermometers are inexpensive and can be very accurate. They do not need a power source. Their main disadvantage is that they are fragile. It would be intolerable to have a thermometer break in the beer. The liquid-in-glass thermometer must be read at the point at which the temperature is measured, which is sometimes inconvenient or hazardous. Liquid-in-glass thermometers are difficult to interface to controllers or data systems.

Dial Thermometer

The sensing element in a dial thermometer is a bimetal strip that consists of two sheets of metal bonded together. When the temperature changes, one metal expands or contracts more than the other, so the strip bends. The strip is wound into a coil so it can move a pointer across a scale (Fig. 10.5).

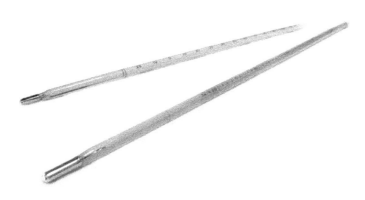

Figure 10.4 Liquid-in-glass thermometers. (See color insert.)

Figure 10.5 Dial thermometer. Courtesy Blichmann Engineering.

Dial thermometers require no power source. They are rugged and can be used in places where there is dust, electrical interference, or vibration. The read-out has to be physically connected to the sensing element. Dial thermometers are not easily integrated into data systems. Accurate dial thermometers can be costly. Bimetal strips can be outfitted with switches that control heating or cooling.

Electronic Thermometer

Electronic thermometers depend on variation of an electrical property of a material with temperature. The signal from an electronic thermometer is an electrical potential (voltage) or current. It is very easy to adapt this signal to a computer or to a digital read-out (a display of numbers) (Fig. 10.6). Another advantage of an electronic thermometer is that the part that measures the temperature does not have to be mechanically attached to the part that gives the reading. A disadvantage is that it needs electrical power.

Remote-Reading Thermometer

A remote-reading thermometer has a bulb containing a liquid or gas connected by a thin tube to a pressure gauge (Fig. 10.7). When the temperature

Figure 10.6 Digital electronic thermometer.

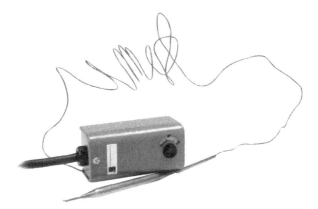

Figure 10.7 Remote-reading thermometer.

increases, the fluid in the bulb expands or vaporizes, causing an increase in the pressure that can be read at the other end of the tube. Remote-reading thermometers are often used as probes to control the temperature in refrigeration systems, such as those used for fermenters. Some advantages of the remote-reading thermometer are that the temperature read-out can be far away from the thing whose temperature is being measured, and that the device does not need electrical power. Disadvantages include the possibility of damage to the long thin tube and limitations on interfacing the signal to other systems.

10.3 COLOR

Light travels as a train of waves. The length of a wave determines its color. The light we can see has waves in the range of 400 to 700 nanometers (16 to 28 millionths of an inch). Beer absorbs short wavelengths that give blue and green colors; the longer waves of yellow and red light can go through to some extent. Darker beers absorb more light and lighter beers absorb less. In the past, beer color was measured by comparison with standard color samples and was reported in degrees **Lovibond** (°L). Pale lager beer might have a darkness of 2 to 4 °L, dark porter beer might have 70 to 100 °L. The inventor of the Lovibond scale, Joseph William Lovibond (1833–1918), founded a company in 1885 called Tintometer, Limited, which still exists. Modern methods for measuring beer color are based on the absorption of light at a particular wavelength, 430 nanometers, in the blue region where beer absorbs light strongly. Two scales are in use. The American Society of Brewing Chemists (ASBC) uses the **Standard Reference Method**, or **SRM**, which gives results that are close to the Lovibond scale. Another scale, called the **European Brewing Convention** (EBC), is used in Europe. The EBC number is roughly twice that of the SRM value. If you take beer that has an SRM color of 10 and mix it with an equal volume of water, the SRM color would be about 5. Both SRM and EBC colors are simplified. They tell how dark the beer is, but not how red or yellow.

Malt is characterized by the color that it imparts to wort. If the malt is labeled 12 °L, it means that wort prepared by a standardized laboratory method with finely ground malt gave a color of 12 °L. The standard wort is prepared with 50 grams of malt and 400 grams of water, which comes to 1.041 pounds per gallon. Brewers mix different malts to give the desired color. Even though malt is often labeled with degrees Lovibond, the measurement may have been made on the SRM scale.

LIGHT AND COLOR: A CLOSER LOOK

The smallest unit of light is called a **photon**. Each photon carries a specific amount of energy that depends on the wavelength of the light that it is a photon of. When a photon encounters a molecule, the photon can be absorbed, that is, the photon transfers its energy to the molecule, or it can be transmitted, that is, it continues on its way. Most of the photons that have a low probability of being absorbed emerge from the sample and give it its color. The molecules in beer absorb blue and green light, so yellow, orange, and red hues emerge to give beer its characteristic amber color.

The more potentially absorbing molecules that a photon encounters as it passes through a sample, the more likely that it will be absorbed. The number of encounters depends on the thickness of the sample and the concentration of absorbing molecules. If a certain sample were to absorb 30% of the light, it would transmit 70%. If we made the sample twice as concentrated, it would transmit 70% of 70%, that is, $0.7 \times 0.7 = 0.49$, which is a transmittance of 49%, so it would absorb 51%. Figure 10.8 illustrates this concept with three identical samples in a row.

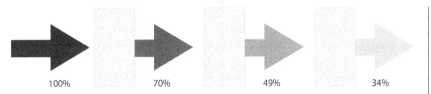

Figure 10.8 Light absorption. (See color insert.)

The quantity that usually follows the concentration in a straightforward way is the **absorbance**, defined by the equation $A = 2 - \log(\%T)$, where %T is the percentage of the light transmitted. If 100% of the light is transmitted, the absorbance would be zero. If 10% is transmitted, the absorbance would be 1. The absorbance varies with the concentration and sample thickness according to an equation called Beer's law: $A = abc$, where a depends on the identity of the sample and the wavelength of the light, b is the sample thickness, and c is the concentration of the light-absorbing material. Very dark or hazy samples do not follow Beer's law precisely.

Both scales of beer color are based on absorbance of light whose wavelength is 430 nanometers (blue). The EBC color is the absorbance of a 10 millimeter thick sample multiplied by 25. The SRM color is the absorbance of a 0.5 inch (12.7 millimeter) thick sample multiplied by 10, or the absorbance of a 10 millimeter sample multiplied by 12.7.

10.4 ALCOHOL CONTENT

There are a number of methods to analyze alcohol. The simplest is to estimate alcohol by volume (ABV) from hydrometer measurements of specific gravity before (OG) and after (FG) fermentation. Table 10.3 gives a calculated value for ABV at 20 °C. There are a number of simplified equations that work in certain ranges.

The most versatile technique to analyze substances that can be vaporized is **gas chromatography**. This method can be used to analyze alcohol and other flavor compounds in beer and, using a different sampling method, to analyze blood alcohol level. As shown in Figure 10.9, a long tube (the column) is packed or coated with a material (the stationary phase) that has a different attraction to the various compounds in the sample. An inert gas like helium, called the carrier, flows through the tube at a constant rate. At the end of the column is a **detector** that measures the presence of anything that is not carrier. The sample is injected into the carrier stream at the beginning of the column. The liquid is rapidly vaporized in a heated chamber called the injector. The gaseous sample travels down the tube with the carrier, but unlike the carrier, the sample molecules stick to the stationary phase to a greater or lesser extent.

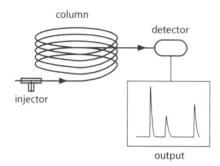

Figure 10.9 Gas–liquid chromatograph.

Molecules in the sample that stick strongly travel slowly. Those that stick weakly or not at all travel quickly. The quickly traveling molecules are separated from the slowly traveling molecules. The different substances in the sample mixture reach the detector at different times. The time it takes a component to reach the detector, called the retention time, is an indication of what the component is, and the response of the detector tells how much of that component is in the sample. Gas chromatography is slow, typically taking 10 to 60 minutes to complete an analysis. It also takes time for the instrument to warm up and come to a steady state.

In large breweries, specialized alcohol analyzers use the absorption of **near infrared** (NIR) light. Near infrared is light that is a little longer in wavelength than visible light. The NIR method is very fast and can be used in a process analyzer. A process analyzer is a device that can be placed in a pipe or tank to provide readings as the material is processed. This contrasts with methods like gas chromatography, which require that a sample be collected and carried to the laboratory for analysis.

ALCOHOL CONTENT FROM SPECIFIC GRAVITY

Until 1995, it was against U.S. law to mention the alcohol content on a beer label. It is now optional, but subject to state and local regulation. If the alcohol content is provided on the label, it must be accurate. Table 10.3 can be used to estimate the alcohol content but should not be used as a substitute for legitimate analysis or for labeling of commercial beer. To read the table, find the row for the original gravity (OG) and the column for the final gravity (FG) in brewer's points. The box where they meet is the estimated alcohol by volume (ABV). For example, a beer whose original gravity is 1.044 (44 points) and whose final gravity is 1.010 (10 points) would have an estimated ABV of 4.46%.

Calculating estimated alcohol content from initial and final specific gravity is surprisingly complex. The basis of the calculation is Karl Balling's classic finding that each gram of dissolved wort solids ferments to 0.4839 gram of ethanol and 0.5161 gram of carbon dioxide and yeast. If we know how much wort solid is fermented (the degree of fermentation), we can determine how much alcohol is made and how much beer remains (net of loss to carbon dioxide and yeast production). We can use this information to estimate the final specific gravity of the beer. This is not perfectly straightforward because the relationships among alcohol content, sugar content, and specific gravity do not follow simple straight lines.

There does not seem to be an elegant method for the reverse calculation, that is, starting with initial and final specific gravity and calculating the degree of fermentation and the alcohol content. Table 10.3 was generated by a computer-based brute force approach in which the degree of fermentation was varied until the correct final specific gravity resulted. It was assumed that the specific gravity is the sum of the contribution from the alcohol and that from the dissolved solids; that the effect of the solids is the same as that of the corresponding mass percent of sucrose; and that Balling's results are applicable. None of these assumptions is perfectly accurate, so the table is no more than it says, an estimate.

TABLE 10.3 Alcohol by Volume Estimation

	FG											
OG	6	8	10	12	14	16	18	20	22	24	26	28	30
30	3.12	2.86	2.60	2.34	2.08	1.82	1.56	1.30	1.04	0.78	0.52	0.26	0.00
32	3.38	3.12	2.86	2.60	2.34	2.08	1.82	1.56	1.30	1.04	0.78	0.52	0.26
34	3.65	3.39	3.13	2.87	2.61	2.34	2.08	1.82	1.56	1.30	1.04	0.78	0.52
36	3.91	3.65	3.39	3.13	2.87	2.61	2.35	2.09	1.83	1.56	1.30	1.04	0.78
38	4.18	3.92	3.66	3.40	3.14	2.87	2.61	2.35	2.09	1.83	1.57	1.31	1.04
40	4.45	4.18	3.92	3.66	3.40	3.14	2.88	2.62	2.36	2.09	1.83	1.57	1.31
42	4.71	4.45	4.19	3.93	3.67	3.40	3.14	2.88	2.62	2.36	2.10	1.83	1.57
44	4.98	4.72	4.46	4.20	3.93	3.67	3.41	3.15	2.88	2.62	2.36	2.10	1.84
46	5.25	4.99	4.72	4.46	4.20	3.94	3.68	3.41	3.15	2.89	2.63	2.36	2.10
48	5.52	5.26	4.99	4.73	4.47	4.21	3.94	3.68	3.41	3.16	2.89	2.63	2.37
50	5.79	5.53	5.26	5.00	4.74	4.48	4.21	3.95	3.69	3.42	3.16	2.90	2.63
52	6.06	5.80	5.53	5.27	5.01	4.74	4.48	4.22	3.95	3.69	3.43	3.16	2.90
54	6.33	6.07	5.80	5.54	5.28	5.01	4.75	4.49	4.22	3.96	3.70	3.43	3.17
56	6.60	6.34	6.07	5.81	5.55	5.28	5.02	4.76	4.49	4.23	3.96	3.70	3.44
58	6.87	6.61	6.34	6.08	5.82	5.55	5.29	5.03	4.76	4.50	4.23	3.97	3.71
60	7.14	6.88	6.62	6.35	6.09	5.82	5.56	5.30	5.03	4.77	4.50	4.24	3.98
62	7.42	7.15	6.89	6.62	6.36	6.10	5.83	5.57	5.30	5.04	4.77	4.51	4.24
64	7.69	7.43	7.16	6.90	6.63	6.37	6.10	5.84	5.58	5.31	5.04	4.78	4.52
66	7.96	7.70	7.44	7.17	6.91	6.64	6.38	6.11	5.85	5.58	5.32	5.05	4.79
68	8.24	7.98	7.71	7.44	7.18	6.92	6.65	6.38	6.12	5.85	5.59	5.32	5.06
70	8.52	8.25	7.98	7.72	7.45	7.19	6.92	6.66	6.39	6.13	5.86	5.60	5.33
72	8.79	8.52	8.26	7.99	7.73	7.46	7.20	6.93	6.67	6.40	6.13	5.87	5.60
74	9.07	8.80	8.54	8.27	8.00	7.74	7.47	7.21	6.94	6.67	6.41	6.14	5.88
76	9.34	9.08	8.81	8.55	8.28	8.01	7.75	7.48	7.22	6.95	6.68	6.42	6.15
78	9.62	9.35	9.09	8.82	8.56	8.29	8.02	7.76	7.49	7.22	6.96	6.69	6.42
80	9.90	9.63	9.36	9.10	8.83	8.57	8.30	8.03	7.77	7.50	7.23	6.96	6.70

Blood Alcohol

In the United States it is against the law to drive a car while one's blood alcohol content is greater than 0.08 gram per 100 milliliters (less for certain classes of driver in some jurisdictions). In law enforcement, breath alcohol is often used as an estimate for blood alcohol. A device that measures breath alcohol in the field is called a **breathalyzer**. Many breathalyzers use an electrochemical method in which alcohol reacts with water at a positive electrode to give acetic acid, hydrogen ions, and electrons: $CH_3CH_2OH + H_2O \rightarrow CH_3COOH + 4H^+ + 4e^-$. The electrons generate an electrical current that can, with careful calibration, be related to the concentration of alcohol present. In addition to their use by the police, breathalyzers are provided by some bars for use by customers, and inexpensive units are available for individual use. Direct testing of blood ethanol is done with a gas–liquid chromatograph (see Fig. 10.9). A vial is partly filled with a blood sample and held at a controlled temperature. A fixed volume of the air above the blood sample in the vial is injected into the chromatograph, which gives a precise measure of blood alcohol.

10.5 pH

Indicators

An indicator is a substance whose color depends on the pH (see Section 4.3). The indicator is a weak acid or base that has different colors in the forms with and without H^+. The reaction can be written

$$HIn(\text{acid color}) \rightarrow H^+ + In^-(\text{base color})$$

where In represents a usually complicated dye structure. Under acidic conditions, the HIn form dominates, so the acid color is displayed. Under basic conditions In^- dominates and the base color is displayed. Different indicators display their acid and base colors at different pH values depending on the pK of the indicator molecules. The pK is the pH at which half of the molecules are in the base form (see Section 4.3, box on "pH: A Closer Look"). When the pH of the solution being tested is lower than the pK of the indicator, the acid color will be displayed. If the pH is higher than the pK, the base color will be displayed. A practical way to measure pH is to use strips of paper or plastic infused with a series of indicators with different pK values. This product is called **pH paper**. The advantage of pH paper is that the pH can be tested with only a drop of sample. Also, no instrumentation apart from the human eye is needed. Calibration (comparison to samples of known pH) is not needed. The disadvantage is that reading pH paper depends on the analyst's judgment and color perception.

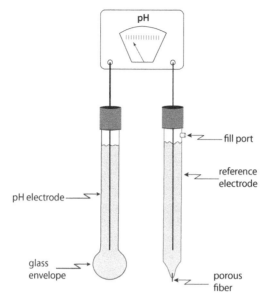

Figure 10.10 pH probe.

pH Meters

A **pH meter** can be used to measure the pH of water-based solutions. The pH meter uses a probe whose electrical potential (voltage) depends on the pH (Fig. 10.10). The signal from the probe is amplified and displayed as a pH. The pH probe uses a device called a **glass electrode**, which has a very thin glass envelope through which only hydrogen ions (H^+) can penetrate. The pH of a solution inside the envelope is held constant. A sensing electrode is immersed in this solution. A reference electrode is in electrical contact with the solution being tested, typically by way of a porous fiber that allows solution from the reference electrode to seep into the sample. For convenience, the two electrodes are often housed in a single device. The difference in the potential between these two electrodes is a measure of the pH of the solution. At room temperature, the potential changes by 0.059 volt per pH unit. The pH meter has the advantages that it can measure pH with far greater precision than indicators, it gives an electrical signal that can be interfaced into a data system, and it gives accurate results even if the analyst is color blind. There are several disadvantages. The reading from a pH meter drifts; it has to be calibrated by comparison to known pH standards at least daily. The meter needs a minimum sample of about 2 ounces (60 mL), which is likely to take several minutes to cool to measurement temperature. The probe is mechanically fragile and susceptible to fouling by mold and bacteria. The glass envelope should not be allowed to dry out; the electrode needs to be stored in a special solution. Even in the best of conditions (which seldom apply to breweries) the electrode needs to be replaced about every two years.

10.6 SENSORY ANALYSIS

There is no scientific instrument that can do the job of human senses. Commercial brewers use groups of tasters known as taste panels to evaluate beer flavor. There are two types of flavor testing, **hedonistic** (Greek: *hedone*, pleasure) and analytical. Hedonistic testing answers the question: "Do I like it?" Large untrained panels of hedonistic testers may be used to develop new products in ways that appeal to a potential market. Analytical testing is designed to evaluate the beer objectively. Trained tasters describe the beer using standard terms like "malty" or "roasty" or "lightstruck." Analytical flavor testing can be used to refine the brewing processes, and to locate and correct flaws in the equipment.

Key issues in analytical testing are avoiding distractions in the testing environment, and making sure that the flavor of the sample is the only information that the taster has available. This is because people, even trained tasters, tend to taste what they expect. For example, if you take pale beer and give it a darker color, testers will tend to report flavors that they associate with dark beer. The tasters may be asked to compare two samples. In another variation, the testers are asked which of three samples is different. Another method is to have the testers rank samples according to a certain characteristic. The results are subjected to statistical analysis to factor out chance variations.

BIBLIOGRAPHY

Anger, Heintz-Michael. Analysis and Quality Control. In *Handbook of Brewing*, H. M. Eßlinger, Ed. Wiley-VCH, 2009, pp. 437–475. Relates degree of fermentation to specific gravity.

Briggs, Dennis E.; Boulton, Chris A.; Brookes, Peter A.; and Stevens, Roger. *Brewing Science and Practice*. CRC Press, 2004, pp. 733–757.

Hackbarth, James J. The Effect of Ethanol–Sucrose Interactions on Specific Gravity. *J. Am. Soc. Brew. Chem.*, **2009**, 67(3): 146–151. Detailed corrections for specific gravity.

QUESTIONS

10.1. For what is a hydrometer used?

10.2. How can temperature be measured?

10.3. What are some advantages and disadvantages of mercury-in-glass thermometers, dial thermometers, and electronic thermometers?

10.4. Describe the operation of a gas chromatograph. Be sure to mention the injector, column, carrier, and detector.

10.5. What is a refractometer? How can one be used in brewing?

10.6. How does a brewer try to control beer color?

10.7. Describe two methods for measuring carbohydrates in beer wort.

10.8. What are the two modern scales for beer color?

10.9. Discuss the difference between hedonistic and analytical flavor evaluation.

10.10. What is a breathalyzer?

***10.11.** What is the carbohydrate content of wort whose specific gravity is 1.058?

***10.12.** Estimate the alcohol content of beer whose OG is 1.048 and whose FG is 1.009.

***10.13.** Estimate the carbohydrate content after fermentation based on the measurements in Question 10.12.

Dried whole hops: the spice of beer. Photo by Mario Zoccoli.

CHAPTER 11

THE CHEMISTRY OF FLAVOR

Flavor involves taste, aroma, and mouth feel. Taste and aroma are chemical senses. The science of flavor is greatly complicated by interactions between the strictly chemical phenomena and other issues, like the visual appearance, order of presentation, and taster expectation. Thousands of papers and hundreds of books have been written about various aspects of flavor. We will have a very superficial look at some of the key issues.

11.1 ANATOMY OF FLAVOR

We learned in Chapter 9 that a living cell is enclosed in a membrane made from a lipid bilayer. The membrane acts as a barrier to ions and polar molecules. Studded about on the surface of the membrane are structures made from one or more protein molecules. Some of these structures, called transmembrane proteins, extend all the way through the membrane. Other structures are attached only to one side of the membrane, either the inside or the outside. Some of these structures are involved in moving ions or molecules into or out of the cells, some transmit signals, and some serve other functions. Structures that act as gates to permit ions or polar molecules to flow into or out of the cell are called **channels**. Structures that drive ions or molecules into or out of the cells from regions of lower to higher concentration (the opposite of the direction of spontaneous flow) are called **pumps**. Pumps need a source of energy to do their work. There are pumps that drive sodium (Na^+) and calcium

---

The Chemistry of Beer: The Science in the Suds, First Edition. Roger Barth.
© 2013 John Wiley & Sons, Inc. Published 2013 by John Wiley & Sons, Inc.

196 THE CHEMISTRY OF FLAVOR

(Ca^{2+}) ions out of the cell and potassium ions (K^+) into the cell, so under normal conditions there is a higher concentration of Ca^{2+} and Na^+ outside and of K^+ inside. Ordinarily the pumps maintain a small negative charge on the inside of the cell, giving rise to an electrical potential (voltage) of about −0.07 volt (−70 millivolts) measured from the outside to the inside of the cell.

Structures that bind to molecules outside the cell and respond with changes inside the cell are called **receptors**. Some receptors are very specific, binding effectively to only a single compound; others, such as those involved with taste and aroma, can respond to a range of compounds. All aromas and some tastes originate with the binding of target molecules to receptors on specialized cells. When the target molecule binds to the receptor, a sequence of events leads to the transmission of a signal to the brain. This signal, integrated with other inputs, is interpreted as an aroma or a flavor.

11.2 TASTE

The sense of taste originates in specialized cells located in small organs in the mouth, mostly on the tongue, called **taste buds**. Taste buds contain taste cells, which are long and thin with one end at the opening of the taste bud into the mouth, and the other end connected to a nerve cell (Fig. 11.1). Different taste cells respond to different taste molecules. There seem to be five distinct tastes: bitter, salty, sour, sweet, and umami (this last being a meaty or broth-like taste). The tastes interact with one another in complex ways. For example, lemon juice alone seems unpleasantly sour, but the sourness is masked in some way when sugar is added to make lemonade. Taste cells respond to taste compounds by changing their internal voltage to a less negative value. This voltage change, called depolarization, leads to the generation of a signal that can be picked up by nerve cells that are connected to the taste cells. The mechanism

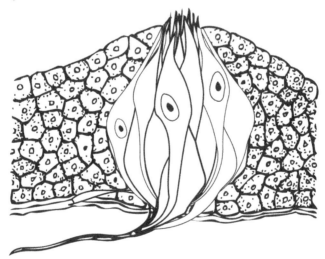

Figure 11.1 Taste bud. Illustration by Marcy Barth.

of this signaling is called **transduction**. The transduction processes for different types of taste cells are different, and the exact mechanisms for some of them are still being investigated.

The salty taste corresponds to ionic compounds with relatively small ions, like sodium chloride (NaCl) and potassium chloride (KCl). These compounds dissolve in the form of ions (for NaCl: Na^+ and Cl^-). Unlike most flavor compounds, they actually enter the taste cells and give rise to an electric signal that is transmitted to the brain and interpreted as a salty taste. Ionic compounds with larger ions, like potassium iodide (KI), give a predominantly bitter taste.

The sour taste is evoked by acidic compounds. These compounds lower the pH by releasing hydrogen ions that combine with water to give the hydronium ion ($H^+ + H_2O \rightarrow H_3O^+$). As with the salty taste, the sour taste cells may respond to the entry of hydronium ions into the cell, rather than relying on binding to surface receptors.

The characteristic sweet compounds are sugars. Nonetheless, a variety of compounds with no evident chemical similarity are sensed as sweet. Many of these are much sweeter than sugar. This makes them useful because sweetness equivalent to a certain amount of sugar can be provided with less compound, and hence fewer calories. For example, aspartame (Fig. 11.2) is about 200 times as sweet as sucrose (table sugar). A gram of aspartame and a gram of table sugar have about the same calories, but the amount of aspartame needed to provide adequate sweetness is so small that its contribution to the calories is negligible. Unlike sugar, many noncarbohydrate sweeteners have perceptible bitterness. This can be a problem because the bitter sensation lasts longer than the sweet sensation, so after a few seconds the predominant sensation is bitterness. No one likes a bitter aftertaste. It is believed that sugars bind to receptor proteins, leading to openings and closings of ion channels giving rise to an electrical signal. Some nonsugar sweeteners may actually penetrate the taste cells and influence the ion channels.

Bitterness is even more chaotic than sweetness. There are thousands of bitter compounds with no evident chemical relationship to one another. The bitter taste is believed to result from the binding of compounds to receptors on the relevant taste cells, leading to changes in ion channels that produce an electrical signal to the brain. People who may be unusually sensitive or insensitive to

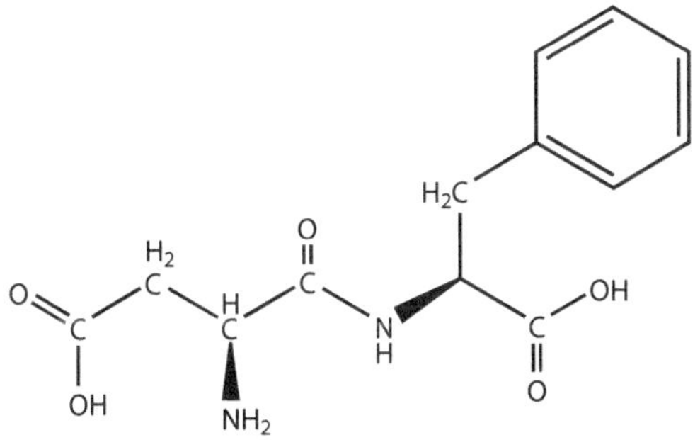

Figure 11.2 Aspartame.

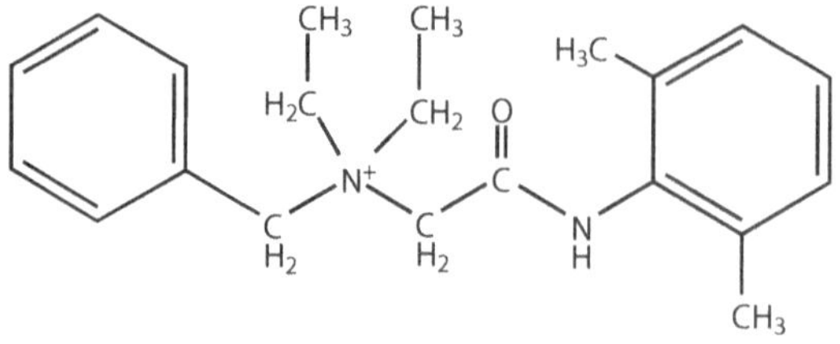

Figure 11.3 Denatonium ion.

certain bitter compounds can be of average sensitivity to others. This suggests that there are several types of bitter receptors. The bitterest species known is the denatonium ion (Fig. 11.3), named for its use in denaturing alcohol.

Some bitterness is desirable in beer. Most of this is provided by iso-alpha acids from hops. We discussed these in Chapter 8 on boiling. The amount of bitterness considered desirable in beer depends on sweet and malty flavors. Sweet or malty beer needs more bitterness to have a balanced flavor. A stout ale could have three times the amount of hop bitter compounds as a standard international lager.

11.3 AROMA

In contrast to the five primary tastes, humans have hundreds of different types of aroma receptors. These interact to produce a nearly unlimited set of aromas. When food is taken into the mouth, vapors can reach the aroma sensors in the nasal lining above the palate. The resulting aroma sensations are combined with the taste sensations in the mouth to give a composite impression identified as flavor. The aroma sensors, in contrast to taste sensors, are actually **neurons** (nerve cells) rather than specialized sense organs attached to nerve cells. These olfactory neurons are connected to a part of the brain called the **olfactory bulb**. No other sensory nerves are connected to the brain by such a direct route. Taken along with taste, the aroma of a substance gives it a specific flavor that we can identify and evaluate.

The exact nature of the aroma receptors and the compounds they bind is the subject of ongoing research. In contrast to color, there is no complete theory of aroma. One reason for this is the complexity of the experiments. Light can be modified instantly with electrical controls. It is easy to show the same thing to any number of subjects. By contrast, each aroma experiment requires a set of aroma samples that are likely to go bad after a few days. Another limitation is that the sense of smell becomes habituated to an aroma; that is, it stops responding to it after a few seconds. The eyes do not exhibit this limitation. There are only three types of sensors for the reception of color. Any color can be perceived from colors detected by these three sensors. If there are primary aromas, we would expect there to be hundreds of them to match the number of receptors. If a typical aroma involved only three of the receptors, there would be millions of possible combinations.

AROMA TRANSDUCTION: A CLOSER LOOK

Despite this complexity, in some ways more is known about aroma than taste. It is generally agreed that aroma-sensing nerve cells have receptor proteins that bind particular aroma compounds at the outside of the cell membrane. This binding produces a change in the part of the protein inside the cell (Fig. 11.4), resulting in the activation of an enzyme that converts adenosine triphosphate (ATP) to cyclic adenosine monophosphate (cAMP), the *second messenger*, and pyrophosphoric

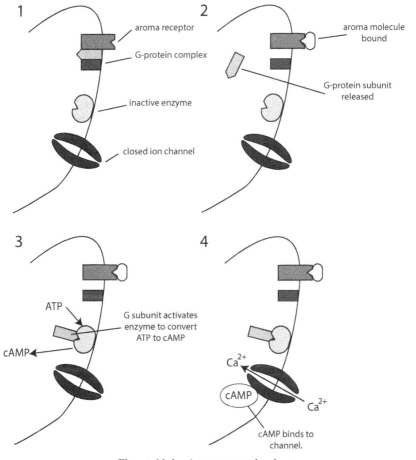

Figure 11.4 Aroma transduction.

(*continued*)

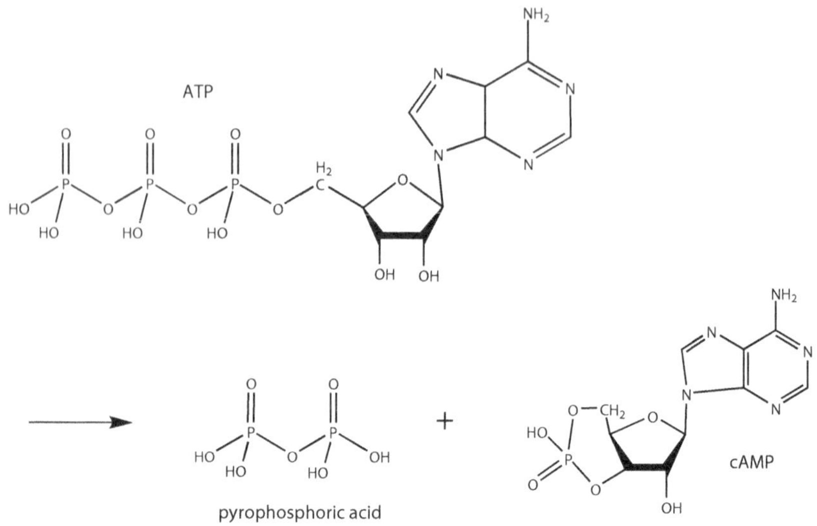

Figure 11.5 Synthesis of cyclic AMP.

acid, $H_4P_2O_7$ (Fig. 11.5). Cyclic AMP binds to specialized calcium ion channels causing them to open and allow calcium ions to flow into the cell. The positive charge of the calcium ions makes the electrical potential inside the cell less negative. If enough calcium ions enter, the potential reaches a value called the **threshold potential**. A potential less negative than the threshold potential causes a large number of **voltage gated channels** to open and admit sodium ions. The cell electrical potential rapidly increases (becomes less negative) going to near zero volts; the nerve cell is said to fire. The signal resulting from the firing of an olfactory neuron is carried to the olfactory bulb of the brain. From there it is transmitted to higher processing centers in the brain, integrated with signals from other olfactory neurons and possibly taste and mouth-feel sensors, and is then interpreted as an aroma or a flavor.

11.4 MOUTH FEEL

In addition to taste buds, the mouth has sensors that react to temperature, pressure, and irritating substances, giving a set of sensory impressions identified as **mouth feel**. Sensations of texture, astringency (drying or puckering), pungency (sharp, peppery), heat, cold, and the prickling sensation produced by carbonation are parts of the mouth-feel experience. Sometimes mouth feel

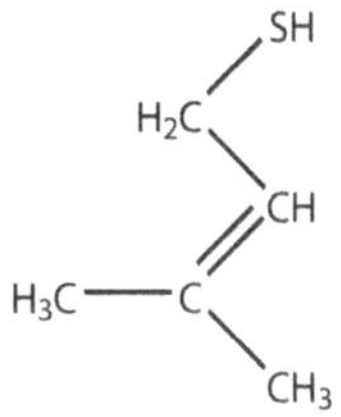

Figure 11.6 3-Methylbut-2-ene-1-thiol (MBT).

is called "trigeminal," a reference to the fifth cranial nerve, called the trigeminal nerve, which carries some of these sensory signals.

11.5 FLAVOR UNITS

For beer flavor compounds, two key aspects are how much is present and how strong its flavor is. Ethanol, for example, is present in beer at about 40 grams per liter (4.0% alcohol by weight). The flavor of ethanol can be detected at a level of 14 grams per liter; this is its **flavor threshold**. So ethanol is present at 2.9 times its threshold (40 g/L ÷ 14 g/L). The ratio of the concentration of a compound to its flavor threshold is the number of **flavor units** of that compound. Ethanol is present at about 2.9 flavor units in average beer. The use of flavor units as a measure of concentration helps us deal with compounds of widely varying flavor thresholds. Some compounds are extremely strongly flavored and have very low thresholds. For example, 3-methylbut-2-ene-1-thiol (MBT, Fig. 11.6), the compound implicated in the odor of lightstruck beer, has a threshold of 7 nanograms per liter (7 billionths of a gram per liter = 7 parts per trillion = 7 ppt). This compound is 2 billion times as strongly flavored as ethanol. The concentration of 3-methylbut-2-ene-1-thiol that would give 2.9 flavor units and would be perceived to give about as much flavor as the ethanol in beer would be about 0.000000002%. Flavor units don't take into account that the perception of flavor is not always straightforward. For most compounds, the perceived flavor increases rapidly above the threshold then slows down; that is, the perceived flavor increases less than the flavor units at higher concentrations. Usually there is a maximum concentration above which there is little change in flavor. Nonetheless, at moderate concentration, the flavor unit is nearly universally used.

Bitterness Units

A bitterness unit (BU), sometimes called an international bitterness unit (IBU), is the outcome of a chemical testing procedure that responds to a variety of hop bittering compounds. The actual perception of bitterness

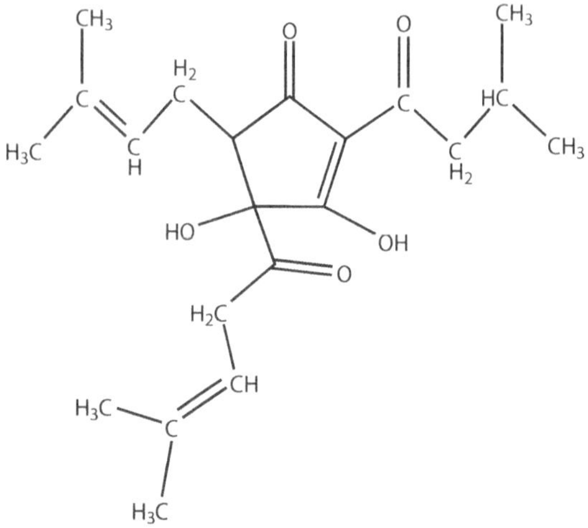

Figure 11.7 Isohumulone.

depends on how much of each compound is present. One bitterness unit is roughly equivalent to one milligram per liter (1 ppm) of isohumulone (Fig. 11.7), the main bitter compound derived from boiling hops. Four bitterness units is equivalent to about one flavor unit. American standard lager has about 10 bitterness units. Bohemian Pilsner and British pale ale have around 30 bitterness units. Dark, malty beers may need higher bitterness levels to balance their maltiness. Guinness Stout is said to have about 45 bitterness units.

11.6 FLAVOR COMPOUNDS IN BEER

Primary Flavor Compounds

Primary flavor compounds are those present at levels higher than two flavor units. For standard lager beer, these include ethanol, isohumulone and related compounds from hops, and carbon dioxide. Some styles of beer have other primary flavor compounds.

Other Flavor Compounds

Secondary flavor compounds, some of which are shown in Figure 11.8, are those present at between 0.5 and 2.0 flavor units. These are the compounds that distinguish one beer from another. Among the more important secondary flavor compounds are esters with fruit flavors, alcohols with more than two carbon atoms (**fusel alcohols**), some organic acids, polyphenols, and peptides (short chains of amino acids). Certain beer styles have hop aroma provided by a variety of compounds from hops. One such compound is myrcene, a

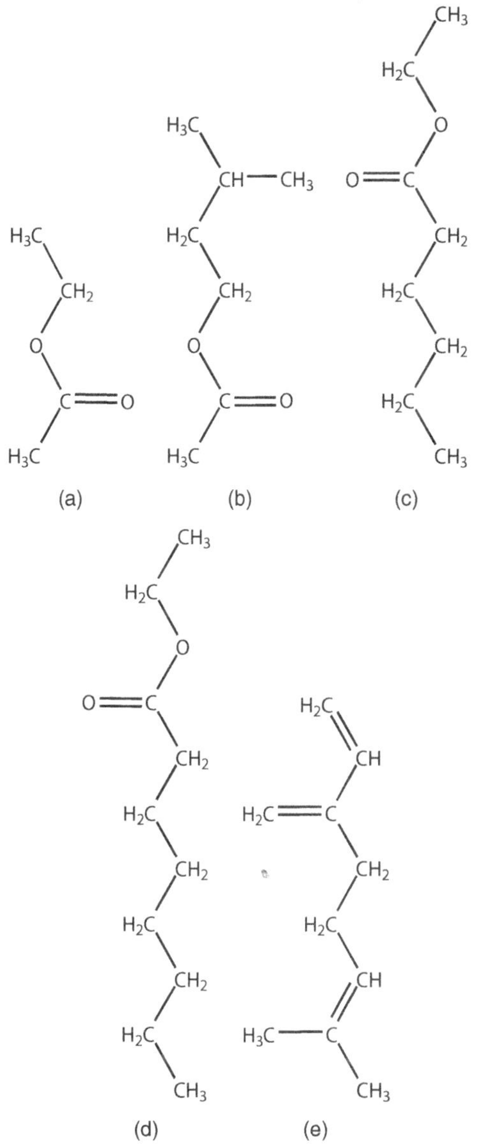

Figure 11.8 Flavor compounds: (a) ethyl acetate (ester), (b) 3-methylbutyl acetate (ester), (c) ethyl hexanoate (ester), (d) ethyl octanoate (ester), and (e) myrcene (terpene).

hydrocarbon (only C and H). Tertiary flavor compounds are those with values between 0.1 and 0.5 flavor unit. The removal of any one of these would not be noticeable, but working together they influence the flavor. Flavor compounds present at levels below 0.1 flavor unit are considered part of the background. (See Table 11.1.)

TABLE 11.1 Flavor Compounds in Beer

Official Name	Common Name	Flavor Units	Character
Carbon dioxide	Carbon dioxide	3.0–5.5	Carbonation
Beer bitter substances	Hop acids	2.0–12	Bitter
Ethanol	Ethyl alcohol	1.4–6.4	Alcohol
Ethyl ethanoate	Ethyl acetate	0.3–1.0	Fruit
3-Methylbutyl ethanoate	Isoamyl acetate	0.1–0.8	Banana
Ethyl hexanoate	Ethyl caproate	0.2–1.3	Apple
Ethyl octanoate	Ethyl caprylate	0.1–0.6	Apple

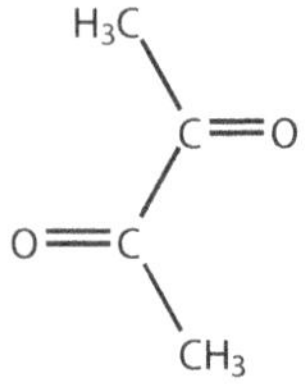

Figure 11.9 Diacetyl.

Specialty Flavors

Certain styles of beer have characteristic flavors that are absent, or nearly so, from standard beer. These include maltiness, phenolic flavor, smoke, and sourness. We will discuss the chemistry of some of these flavors in the next chapter on the chemistry of beer styles.

Off-flavors

In practice, an **off-flavor** is one that does not meet the consumers' expectations. There are many stories of commercial beers with long standing defects that became part of the expected profile of the beer. It is likely that some of these stories are actually true. It is demonstrably true that a flavor that is considered to be a defining characteristic of one style would be regarded as a serious defect in another. This makes sense. The flavors that define lambic ale do not belong in Continental lager. There may be off-flavors present in beer when it is brewed, and there are off-flavors that develop as the beer ages. We will look at just a few sample compounds.

Diacetyl (Fig. 11.9), officially called butanedione, gives a buttery flavor that is considered seriously nasty in pale lager beer. It is used as an artificial flavoring in popcorn. Diacetyl has a flavor threshold of 0.15 milligram per liter (0.15 ppm). It is made by yeast during fermentation and should be consumed by the yeast during conditioning. Excessive diacetyl can result from errors in conditioning the beer, but the usual problem is unhealthy yeast or contamination by wild yeast or *Pediococcus* bacteria.

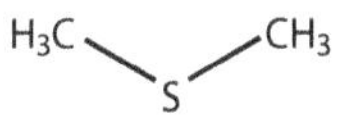

Figure 11.10 Dimethyl sulfide.

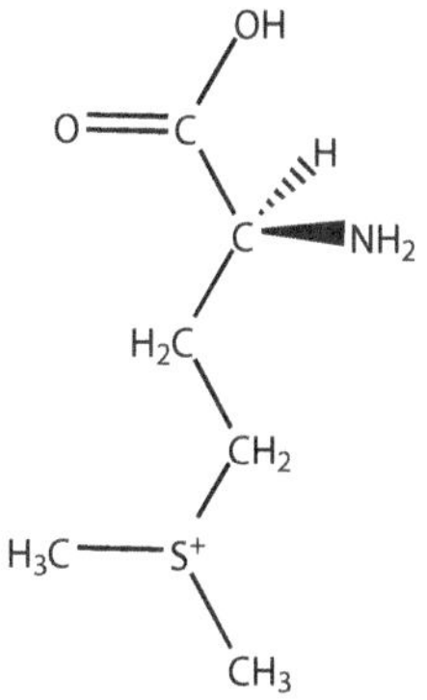

Figure 11.11 S-methyl methionine.

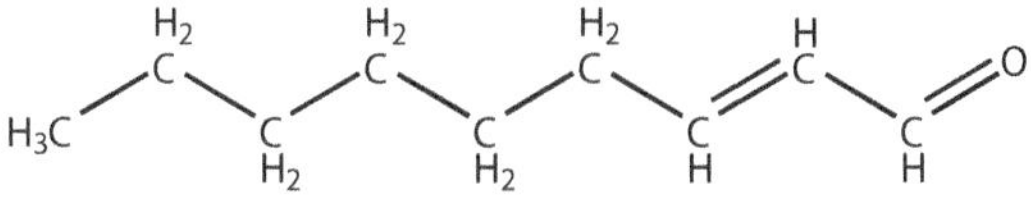

Figure 11.12 Trans-2-nonenal.

Dimethyl sulfide (DMS, Fig. 11.10) can give a cooked vegetable flavor to beer. It has a flavor threshold of 50 micrograms per liter (50 ppb). Dimethyl sulfide is continuously produced in hot wort from a sulfur-containing amino acid called **S-methylmethionine** (Fig. 11.11) from the malt. During boiling DMS goes off with the steam. DMS can be present at objectionable levels if boiling is not vigorous enough, if condensed steam is allowed to return to the kettle, or if some of the wort stays hot for a long time after boiling stops. This is the reason that wort is chilled as quickly as possible after the boil.

Trans-2-nonenal (Fig. 11.12) gives stale beer a cardboard flavor. The aroma threshold for this compound is 0.11 microgram per liter (0.11 ppb), so it doesn't take much. Several other compounds are implicated in staling of beer. The reactions producing these compounds are faster at higher temperatures, so beer should be stored in a cool place. Part of the problem is that the brewer has little control over what happens to the product after it leaves the brewery. Beer staling is discussed in Chapter 15.

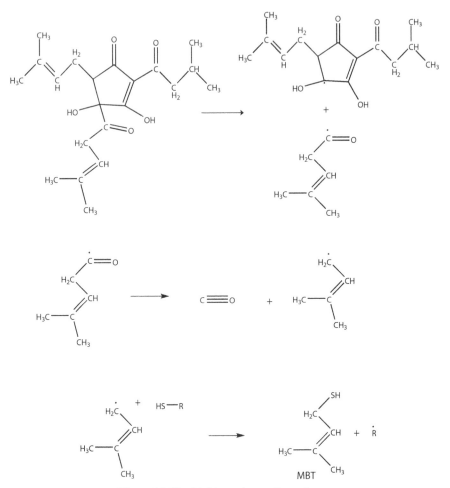

Figure 11.13 Lightstruck reaction scheme.

Beer that has been exposed to light can develop a foul flavor reminiscent of the odor of a skunk. The beer is said to be **lightstruck**. This term was selected in favor of the more familiar (to Americans) "skunky" because there are no skunks in Europe. Only beer with hops gets lightstruck. The reactions leading to lightstruck flavor are shown in Figure 11.13.

Energy from the light is transferred to the hop-derived iso-alpha acids. This energy breaks the bond of a six-carbon side chain, leaving one of the shared electrons (shown as dots) on each of the two broken parts of the molecule.

These smaller molecules are unstable **free radicals** (molecules with unpaired electrons; most are very reactive). The smaller part releases carbon monoxide (CO), leaving an unstable five-carbon free radical. This free radical reacts with sulfur-containing proteins (shown in the diagram as H–S–R) to give 3-methylbut-2-ene-1-thiol (MBT), which has an odor like skunk spray. It is worth noting that MBT is not a component of actual skunk spray.

MBT has a threshold of about 7 nanograms per liter (7 ppt). The usual defense against lightstruck beer is to keep the beer out of light. Brown bottles let in very little light at the wavelength that can lightstrike beer; cans and kegs let in none. Brewers whose marketing managers insist on colorless or green bottles can use hydrogenated hop extract instead of natural hops. These products are treated (see Fig. 8.6) to eliminate one or more of the double bonds on the side chains of the iso-alpha acids, a process called "skunk-proofing." Skunk-proofing, depending on which double bonds are eliminated, can prevent light from interacting with the bittering compounds. Alternatively, light may break the bond to the side chain, but the product, which lacks a double bond, is much less strongly flavored than MBT, so small amounts are below the flavor threshold.

BIBLIOGRAPHY

Buck, Linda; and Axel, Richard. A Novel Multigene Family May Encode Odorant Receptors. *Cell*, **1991**, *65*(1): 175–187. Reports the discovery of the genes for olfactory receptors. Buck and Axel won the Nobel Prize in Medicine or Physiology for this work in 2004.

Burns, Colin S.; Heyerick, Arne; De Keukeleire, Denis; and Forbes, Malcolm D. E. Mechanism for Formation of the Lightstruck Flavor in Beer Revealed by Time-Resolved Electron Paramagnetic Resonance. *Chem. Eur. J.*, **2001**, *7*(21): 4553–4561. How beer gets lightstruck.

De Kuekeleire, Denis; Heyerick, Arne; Huvaere, Kevin; and Skibsted, Leif H. Beer Lightstruck Flavor: The Full Story. *Cerevisia*, **2008**, *33*(3): 133–144. It's the full story.

Firestein, Stuart. How the Olfactory System Makes Sense of Scents. *Nature*, **2001**, *413*: 211–218. A readable review of how the olfactory system works.

Margolskee, Robert F. Molecular Mechanisms of Bitter and Sweet Taste Transduction. *J. Biol. Chem.*, **2002**, *277*(1): 1–4. A mini-review of how bitter and sweet receptors work.

Meilgaard, Morton C. Flavor Chemistry of Beer: Part II. *MBAA Tech. Quarterly*, **1975**, *12*(3): 151–168. Seminal paper on flavor thresholds of beer compounds.

Meilgaard, Morton C. The Flavor of Beer. *MBAA Tech. Quarterly*, **1991**, *28*: 132–141. A review of aspects of flavor and flavor evaluation.

Wood, William F. A History of Skunk Defensive Secretion Research. *Chem. Educator*, **1999**, *4*(2): 44–50. A review of what is known about skunk spray.

THE CHEMISTRY OF FLAVOR

QUESTIONS

11.1. What is the difference between taste and flavor?

11.2. Describe a lipid bilayer. What is its biological function?

11.3. What are the functions of channels and pumps?

11.4. If a certain compound has a flavor threshold of 7 ppm, how many flavor units would it have in a sample in which its concentration is 21 ppm?

11.5. A certain homebrewer uses no chiller. He allows the wort to cool naturally from the boiling temperature over a period of 24 hours. What defect is likely to be present in the beer?

11.6. Identify the flavors associated with the following compounds.

(f)

(g)

Beer aging barrels. Boston Brewing Company, Boston, MA.

CHAPTER 12

THE CHEMISTRY OF BEER STYLES

A beer style originates when several brewers in a locality begin to produce a distinct variety of beer. The interactions among the nature, quality, and cost of locally available brewing supplies (water, grain, fuel, adjuncts, flavorings, and yeast), taxes and regulations, transportation issues, customer preferences, and brewing traditions determine the nature of the style. Most of the styles of beer we drink today originated in the late Middle Ages or later. This makes them recent innovations compared to the long history of brewing. There are dozens of accepted beer styles and new styles are developed and compete for acceptance regularly. Modern communication and mobility have removed many restrictions to the free movement of people, goods, and ideas, so the local aspect of beer styles is less relevant. A style made famous in Belgium may see its most authentic realization in Italy, California, or Mexico.

The characteristics of a beer style involve flavor, strength, and appearance (color, head). These characteristics are achieved by the types and amounts of ingredients and by the processing conditions. The variables interact in complex ways, so we will limit our discussion to a few basic generalities. We will look at the characteristics of some broad families of styles. We will then describe a particular classic beer style as an example of how important style characteristics are brought about.

12.1 BEER STYLE FAMILIES

There is no universal way to classify the large number of beer styles that are available. Most classifications go by color and by the country or town in which

The Chemistry of Beer: The Science in the Suds, First Edition. Roger Barth.
© 2013 John Wiley & Sons, Inc. Published 2013 by John Wiley & Sons, Inc.

the style originated. These schemes are not helpful in considering the chemistry of the beer, so we will try to sort beer styles by flavor characteristics. The following scheme is essentially arbitrary and is intended to allow us to look at a big picture at the expense of many important details. We will start with the two major divisions of beer styles, ale and lager. Ales usually have fruity flavors contributed by esters formed during fermentation. Lagers usually do not have as great a concentration of esters because of the yeast strains and low fermentation temperatures used. Within this broad division we will classify styles largely on the basis of whether they emphasize bitterness or maltiness. We will make two additional distinctions to account for the characteristic phenolic flavor of wheat ales, and a variety of special flavors.

Bitter Ales

This style family includes styles like Pale Ale, Bitter, Saison, and others. The color tends to be moderate, in the range of 5–15 SRM. The bitterness ranges from the high side of moderate to very intense, especially for some American styles. Maltiness is generally low to moderate. The base malt is usually pale ale malt, often with some crystal malt added. Hop aroma varies but is seldom totally absent.

It is interesting that pale ale malts and, in fact, all light-colored malts could only be produced in quantity when materials and techniques for indirect heating of the malted barley had been developed and scaled up. Before technological advances in the use of iron for furnaces and coal for fuel, the usual method to dry malt was to put it in a room with a wood or peat fire. The malt absorbed smoke from the fire, which gave it a smoky color and flavor. The depletion of forests led to the use of coal for fuel. Coal has a significant sulfur component, which can impart bad flavors to malt dried with it. To get an acceptable product the heat had to be applied indirectly so that the smoke from the coal would not touch the malt. The new fuel led to a new malt, which led to a new family of beer styles.

Malty Ales

Ales whose flavor is dominated by maltiness fall into the category of malty ales. Because the malty flavors generally arise from heat treatments that also make the beer darker, most malty ales are dark. Some examples are brown ale, red ale, porter, stout, and Scottish ales.

The browning reaction sequence that is most important in brewing is called the **Maillard** (MAY yard) **reaction**. The first step is a reaction between a sugar molecule and an amino acid to give a Schiff base, shown in Figure 12.1. The Schiff base can react in many ways, some of which lead to large, highly colored molecules called **melanoidins** and some of which lead to smaller highly flavored molecules, like furaneol (4-hydroxy-2,5-dimethyl-3-furanone, also called DMHF) and maltol, shown in Figure 12.2. This is why darker foods tend to be

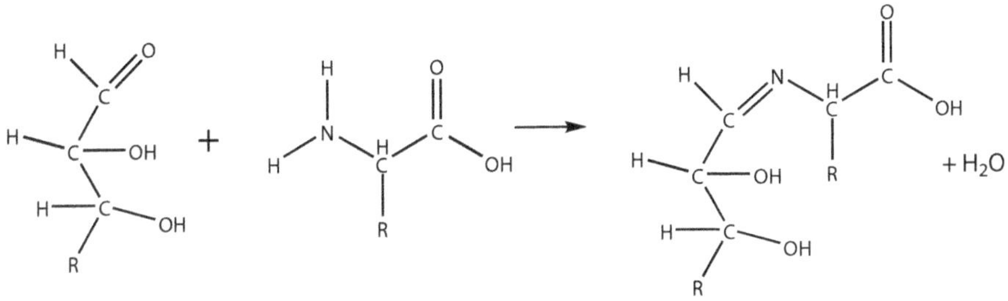

Figure 12.1 Maillard reaction first step: carbohydrate + amino acid → Schiff base.

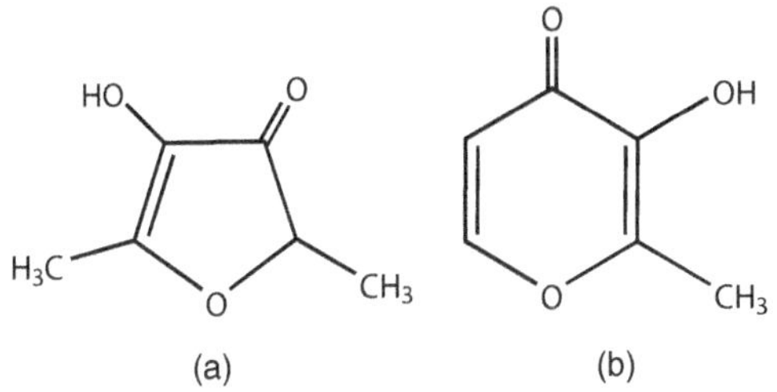

Figure 12.2 Maillard products: (a) furaneol and (b) maltol.

more highly flavored. Some Maillard flavor molecules, like furaneol and maltol, have nut-like, toasted, or malty flavors characteristic of dark beers. The Maillard reactions go most rapidly at high temperature and under dry conditions.

The ideal conditions for the Maillard reaction occur during high temperature roasting of grain (malted or not). Most brewing malt is dried at moderate temperature. If a darker, more highly flavored malt is desired, the dry malt is then roasted. Roasted malts are used to give characteristic flavors and colors to the styles of beer that use them. Another family of malts is heated at the saccharification temperature of 149–167 °F (65–75 °C) before drying, a process called "stewing." Stewing converts most of the starch to sugar. The resulting malts are variously called *crystal* or *caramel* malts. After stewing, the malt is subjected to drying and kilning that can vary in duration and temperature, giving grades of malt with a series of colors and flavor levels.

Phenolic Ales

The phenolic flavor is often described as resembling cloves. It is characteristic of some wheat-based ales and a number of Belgian barley-based ale styles. The phenolic flavor is considered a defect in other styles, which is why it is often called phenolic off-flavor (POF). The major contributor to the phenolic flavor is 4-vinylguaiacol, which is produced by the removal of carbon dioxide from ferulic acid, a component of plant cell walls (Fig. 12.3). Ferulic acid is

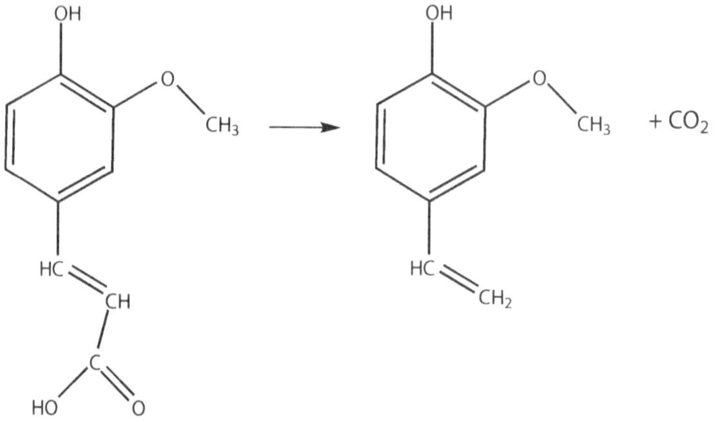

Figure 12.3 Phenolic flavor: ferulic acid → 4-vinylguaiacol + CO_2.

present in barley and wheat malt at levels well below its flavor threshold of 600 milligrams per liter. The conversion of ferulic acid to 4-vinylguaiacol and carbon dioxide occurs to some extent during mashing and boiling, but most of it is usually formed by the yeast during fermentation. This is significant because the flavor threshold of 4-vinylguaiacol is a factor of 2000 lower than that of ferulic acid. Small traces of 4-vinylguaiacol can dominate the flavor of beer. Certain strains of yeast, which are said to be POF+, produce an enzyme that catalyzes the reaction. Phenolic ale styles that use POF+ strains for fermentation can have 4-vinylguaiacol concentrations above 2 flavor units.

Bitter Lagers

Many lager beers have barely discernible maltiness; their flavor is dominated by hop bitterness. Some bitter lager styles include standard American lager, Pilsner lager, and Kolsch.

Malty Lagers

Some malty lager styles include Munich Dunkel, Munich Helles (a pale, but malty lager), and the various types of Bock beers.

Specialty Beers

This category includes ales and lagers that have special flavors. These include spiced beers, such as pumpkin ale, smoked beer, sour beer, fruit beer, jalapeño pepper beer, and a myriad of other products of the fertile imaginations of brewers. The most familiar sour style is lambic ale, in which fermentation conditions invite a number of species of yeast and bacteria to produce a complex flavor profile. Sourness usually comes from lactic acid (Fig. 12.4).

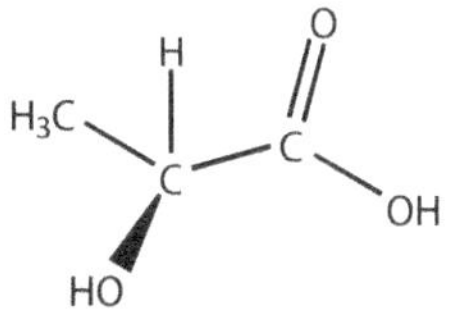

Figure 12.4 L-Lactic acid.

12.2 REALIZING A STYLE

We will now use Bock beer as an example of how a style is realized in the brewery. Bock beer as we know it today is a 17th century Munich revival of a style developed hundreds of years earlier in Einbeck, a town in north-central Germany. Modern Bock is a lager with an original gravity around 1.067, a final gravity of about 1.016, about 25 bitterness units, and a color of 17 SRM. It has a very malty flavor, no hop aroma, and no fruity or buttery notes. The mouth feel is full or thick; carbonation is low. We will use this description as our example as we consider the various characteristics of a beer style and how they arise from the chemistry of the malting and brewing processes.

Original Gravity

The original gravity is the specific gravity of the wort before fermentation begins. It is one of the key defining characteristics of a beer style. Specific gravity is used as a measure of the carbohydrate content of the wort. Other things being equal, a higher carbohydrate content before fermentation leads to a higher alcohol content after fermentation. There are several measures of carbohydrate content, including the specific gravity. Among these are weight percent, pounds of carbohydrate per gallon of liquid, and grams of carbohydrate per liter of liquid. The weight percent is the most general.

$$\text{wt}\% = \frac{\text{mass carb}}{\text{total mass}} \times 100\%$$

The weight percent is essentially equivalent to the degrees Plato (°P), a measure often used in the brewing industry. Sometimes it will be convenient to use the weight fraction (decimal equivalent of the weight percent), that is, the weight percent divided by 100%. Specific gravity is often expressed in brewers points. A specific gravity of 1.067 would be 67 brewer's points. To approximate the weight percent from the brewers points for specific gravities up to 1.050 (50 points), divide the points by 4. Above 50 points, divide by 4.1 (this is because the specific gravity is not a perfectly linear function of sugar content). For Bock at 67 points, the weight percent carbohydrate is about 67 divided by 4.1, which comes to 16.3%. Table 10.2 gives the more precise value

TABLE 12.1 Typical Potential Extract and Color

Malt Type	Potential Extract (pt/lb/gal)	Potential Extract (%)	SRM Color
Pilsner	37	81	2
Pale Ale	36	80	3
Vienna	36	79	4
Munich	36	79	6
Dark Munich	36	78	9
Biscuit	36	77	25
Chocolate	35	75	350
Black	34	73	510
Crystal	34	73	Various

Source: Maltster specifications.

of 16.37%. That means that 100 ounces (by weight) of wort would contain 16.37 ounces of carbohydrate (or 100 pounds would contain 16.37 pounds, or 100 grams would contain 16.37 grams).

The original gravity is set by the amount of water, starchy ingredients, the efficiency of mashing, and any added carbohydrate sources like sugars or malt extract. The amount of carbohydrate provided by each type of starch source is available in published tables. Using this information and some seriously messy calculations (see box on "Computing Recipes"), the brewer can devise a recipe that will give the required volume of wort of the required original gravity (see Table 12.1).

Color

Beer color derives from a family of compounds called melanoidins. These are complicated and poorly characterized molecules that result from the Maillard reaction (see Fig. 12.1). Melanoidins are most readily formed at elevated temperature and in a dry environment, but they can form more slowly under moderate temperature and in wet conditions. The conditions that produce melanoidins also give compounds with roasted or malty flavors, so dark colors and more intense flavors often go together in beer and other foods that are browned in heat. Melanoidins are produced during drying, kilning, and roasting of malt, and during high temperature mashing and boiling.

Malts range in color from the pale yellow Pilsner malt to the intensely colored black patent malt. There are also several grades of crystal malt ranging in color from 10 to 160 SRM. Crystal malt is produced by heating the malt without drying. This saccharifies much of the starch in the malt grains and gives caramel flavors. The "malt color" is the SRM or EBC color (see Section 10.3) of wort produced by mashing 50 grams of finely ground malt with 400 grams of distilled water under laboratory conditions. This comes to 1.04 pounds of malt per gallon of water. How this translates to actual brewing conditions is not clear. We can guess that the color would follow the brewhouse efficiency, that is, if the brewer gets 75% of the laboratory value for extract, he or she would get about the same 75% of the laboratory color. The pounds per gallon of a certain malt times the malt color (SRM or EBC), divided by 1.04, and multiplied by the brewhouse efficiency will be our estimate of the color contribution for that malt.

For many types of brewing, the color is set by the grain; the darkening produced by boiling is a small adjustment. The exceptions are for very pale beers and some styles that call for multiple boiling steps during the mash. A reasonable starting estimate for the darkening produced by a 60 minute boil would be 2 SRM. If the boiling is longer, or if the mash is decocted (see Section 7.4), the beer will be darker. For many styles, the bulk of the malt is pale with a small amount of crystal or roasted (chocolate or black) malt making a large contribution to the color. Most grain adjuncts like rice or corn grits have a color potential between 1 and 2 SRM. Sugars like cane, beet, or corn sugar have zero SRM, but some heat-treated sugar products, like dark Belgian Candi sugar, can be as dark as 275 SRM. The very light colors of some brands of beer are achieved by the heavy use of pale adjuncts with pale malt.

COMPUTING RECIPES: A CLOSER LOOK

Under ideal laboratory conditions, about 70–80% of most types of malt can be extracted into the wort. Dry malt extract and sugar have an extract potential of nearly 100% (it would be 100% if the solid were perfectly dry). To simplify the calculation, we will convert the potential extract from weight percent, as found in maltsters' specifications, to points per pound per gallon (pppg). This is the number of specific gravity points for each pound of malt per gallon of water (see Table 12.1). In the range of 70–80% extract, the pppg is the extract percent times 0.463. Outside of the laboratory, the malts do not yield this much carbohydrate. If we know that we can only get 75% (the **brewhouse efficiency**) of what is available, we would have to multiply the potential extracts by 0.75 (the decimal equivalent of 75%) to get the actual yields. Sugars and malt extracts are not subject to losses in mashing, so their brewhouse efficiencies are 100%. Suppose that, for reasons of flavor and color, we decide to use 15% crystal malt with an extract potential of 34 pppg, 10% pilsner malt with a potential of 37 pppg, and 75% Munich malt, with a potential of 36 pppg. For each item we calculate points from the mass times the potential extract (PE). We multiply this by the brewhouse efficiency, 0.75 to get the actual points. Consider a hypothetical batch with 100 pounds

(*continued*)

of grist. We can set up a table showing how many points we expect from each type of malt.

Malt	PE	Total	Potential Points	Actual Points
Crystal	34	15 lb	510	383
Pilsner	37	10	370	278
Munich	36	75	2700	2025
Total		100	3580	2686

The total amount of carbohydrate from 100 pounds of grist would be 2686 points, so each pound provides 26.86 points. Suppose we intend to make 5.5 gallons of wort (a 5 gallon batch with allowance for some left behind in the yeast and **trub**). The desired original gravity is 1.067 or 67 specific gravity points per gallon. The total points is calculated from the volume times the specific gravity points per gallon, which is 67 × 5.5 gal = 369 points. Each pound of grist provides 26.86 points, so we need 369 points divided by 26.86 points per pound = 13.7 pounds. The recipe is shown below.

Crystal	15%	2.0 lb
Pilsner	10%	1.4
Munich	75%	10.3

Styles, Color, and Water

Darker colored malt gives a more acidic mash than paler malt. This is because the Maillard reaction (see Fig. 12.1) removes amines, which are basic compounds, allowing the mash pH to be lower. Over the centuries, brewers whose water is highly alkaline (Section 4.4) learned that they could make better beer if they used darker malt. We now know that the acidity of the dark malt helps to counteract the alkalinity of the water, bringing the pH to an acceptable level. The characteristic styles in alkaline water regions, like Dublin and Edinburgh, are dark and malty. Areas with water of low alkalinity, like Pilsen, specialized in pale beer. Hardness increases mash acidity, favoring paler malt and paler beer styles. Sulfate sharpens hop bitterness; beer brewed in high sulfate areas tends to be bitter. Perhaps the most extreme water is that of Burton upon Trent, the beer capital of England. Burton water is very hard, moderately alkaline, and has a very high concentration of sulfate ion. The effects of the hardness and alkalinity cancel one another to some extent (Section 4.4, box on "Residual Alkalinity"). Burton is famous for bitter pale ale.

COLOR CALCULATION: A CLOSER LOOK

The color of our Bock beer is supposed to be 17 SRM. We can adjust the color by using two colors of crystal malt adding up to 2 pounds total, to give the correct gravity. We adjust the amounts of the darker and lighter malt to give the desired color.

To simplify the calculations we will determine the total Malt Color Units (MCU) contributed by each type of malt. The lab MCU for each type of malt is the weight of the malt in pounds divided by 1.04 times the SRM malt color. The adjusted MCU is the lab MCU times the brewhouse efficiency expressed in decimal form. The net color will be the sum of the adjusted MCU of each malt divided by the volume of the batch. We will use a brewhouse efficiency of 75% = 0.75.

Malt	Color	Mass	Lab MCU	Adjusted MCU
Crystal 20	20	1.4 lb	27	20
Crystal 40	40	0.6	23	17
Pilsner	2	1.4	3	2
Munich	6	10	58	43

The total MCUs add up to 82. Dividing by the batch volume 5.5 gallons gives a color of 15 SRM. If the wort darkens by 2 SRM during boiling, we will be very close to the target color.

Bitterness

Bitterness in beer derives mostly from the isomerization of alpha acids from hops. This process occurs during the boil. The more alpha acids are present, and the longer they are boiled, the more bitter the beer will be. A secondary, but significant effect is that the isomerization process is slower when the wort gravity is higher. Some bitterness is lost during fermentation. Bitterness declines with time after the beer is packaged. The fraction of the alpha acids isomerized and providing bitterness in the final beer is called the hop utilization. For a 60 minute boil and an original gravity of 1.040, a reasonable starting estimate would be 25% utilization. Our Bock has an original gravity of 1.067, so we will estimate the utilization at 20%.

The target bitterness is 25 bitterness units, which is equivalent to about 25 milligrams of iso-alpha acid per liter of beer. A 5.5 gallon batch comes to 21 liters, so we need 25 milligrams per liter times 21 liters = 525 milligrams. We divide the iso-alpha acid requirement by the decimal equivalent of the utilization: 525 ÷ 0.20 = 2625 milligrams = 2.63 grams. So we need enough hops to provide 2.63 grams (0.0928 ounce) of alpha acids. Typical hops used for Bock would be Hallertau. Suppose the Hallertau has an alpha acid content of 4.0% (information provided by the supplier). To get the mass of hops needed we divide the alpha acid requirement by the decimal equivalent of the alpha acid content: 0.0928 ounce ÷ 0.040 = 2.32 ounces.

Flavor

The intense malty flavor in Bock beer is traditionally provided by Munich malt and by browning reactions during decoction mashing and during a long kettle boil. The grist is mashed in with cold water and brought to a lukewarm temperature with boiling water. One-third of the mash is put into the kettle and heated first to the saccharification temperature, then to boiling for 10–30 minutes. The boiling mash is returned to the mash tun, raising the temperature of the mash to the lower end of the saccharification range. This process is repeated twice more, each time raising the mash temperature to the next step. After mashing and wort separation, the entire wort is boiled for 90 to 180 minutes. These prolonged boiling steps make the beer darker and add toasted and malty flavors because of Maillard (browning) reactions between the proteins and the carbohydrates. Boiling the wort to produce browning products ties up the equipment for a long time, limits output, and consumes a large amount of energy.

Crystal malt has been used to produce malty flavors in Bock beer instead of the traditional decoction mashing and boiling procedures. Some regard this practice as a responsible way to conserve resources while still producing an acceptable product. Others question the authenticity and acceptability of Bock produced in this way. Fortunately, the market has room for both methods and consumers can make their own choices.

The thick, full mouth feel of Bock comes from dextrins. Dextrins are soluble but unfermentable carbohydrates with chains of more than three glucose units. In a simplified model, we could say that dextrins are made by the action of alpha-amylase, which hydrolyzes starch molecules in the middle of the long chains. Dextrins are removed by beta-amylase, which hydrolyzes near the ends of the dextrin chains to give maltose. By selecting a mashing program that avoids the temperature range in which both alpha- and beta-amylase are very active, the dextrin concentration can be enhanced. The brewer must not go too far in this direction; dextrins do not ferment to give alcohol. Bock is supposed to be a strong beer.

The other mouth feel issue is carbonation level, that is, the concentration of carbon dioxide in the beer. The most common measure of carbon dioxide is the volume of carbon dioxide per volume of beer, measured at 0 °C (32 °F) and atmospheric pressure. Each volume of carbon dioxide per volume of beer comes to 0.2% CO_2 by weight. Bock is carbonated at about 2.2 volumes of CO_2 per volume of beer. This compares to 2.7 for other European lager beers and 2.0 for British ales. The solubility of a gas in a liquid depends on the pressure of the gas and the temperature of the liquid. The pressure dependence follows Henry's law: $s = k \cdot P$, where s is the solubility, P is the pressure, and k depends on the temperature of the liquid in a complicated way, but always becomes lower as the temperature increases. To reach a certain level of carbonation, the brewer could hold the beer at a particular temperature and admit carbon dioxide at the pressure corresponding to the desired level of carbonation. For example, at 41 °F (5 °C) the carbonation will reach 2.2 volumes per volume of beer at an applied gauge pressure of 10 pounds per square inch (10 psig). If the temperature were higher, a higher pressure would be needed. In practice it takes a long time (days) for the carbon dioxide to dissolve if the minimum pressure is used, so the brewer will often apply a higher pressure and carefully monitor the total amount of carbon dioxide admitted so the correct carbonation level can be achieved in a few minutes.

Other flavor issues in Bock beer are largely concerned with the absence of certain flavors. Bock beer has no perceptible hop aroma. A hop variety with a low to moderate oil content is selected. The long boiling helps to assure that any aroma compounds from the hops are boiled off.

Beer styles that are supposed to have hop aroma, like American Pale Ale, must be treated in ways that preserve these compounds (Fig. 12.5). If a portion of the hops is added shortly before the wort is chilled (late hopping) or even after the wort is chilled (dry hopping), hop aroma compounds can be carried into the finished beer. Another method to introduce hop aroma is to run the hot wort from the kettle through a bed of hops in a device called a hop back. The wort then goes either directly or via the whirlpool to the chiller. Because there is no place for the volatile hop aroma compounds to go, they are retained in the beer. These late hop additions add little bitterness because there is not much time at the boiling temperature for alpha acid isomerization to occur. Different varieties of hops have different aroma compounds, so the brewer must select varieties that give the desired aroma.

Bock has no fruity, flowery, or buttery flavors. Fruity or flowery flavors can come from the hops or from esters (Fig. 12.6) produced by the yeast during fermentation. Ester production can be minimized by selecting a lager yeast

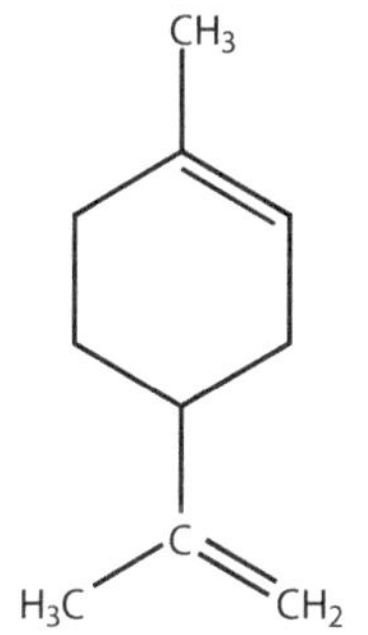

Figure 12.5 Limonene (hop aroma).

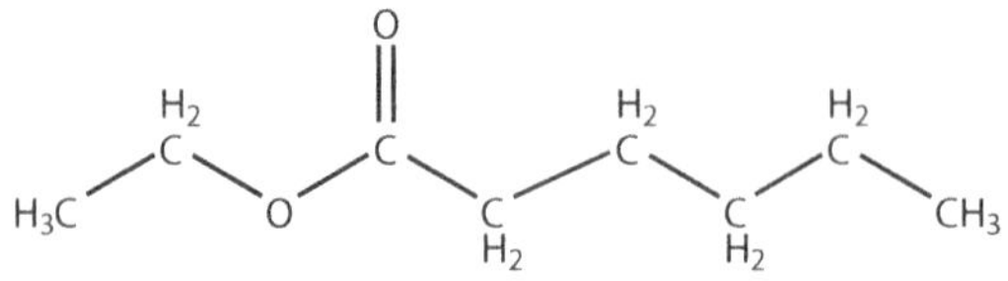

Figure 12.6 Ethyl hexanoate (ester).

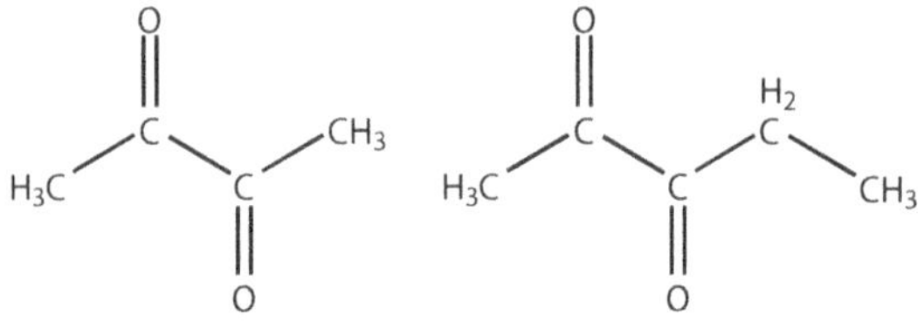

Figure 12.7 Vicinal diketones: (a) diacetyl and (b) 2,3-pentanedione.

strain with a reputation for low esters and by fermenting at a low temperature, in the 48–55 °F (9–13 °C) range. Low temperature fermentation has a tendency to increase the concentration of diacetyl and 2,3-pentanedione (collectively called vicinal diketones, Fig. 12.7), which have an objectionable buttery flavor. The yeast will consume the vicinal diketones at the end of the fermentation, but a prolonged low temperature conditioning period is needed to assure compete absence of any buttery flavors.

BIBLIOGRAPHY

American Society of Brewing Chemists. *Methods of Analysis*, 7th ed. 1976. How everything is measured in the United States.

Coghe, Stefan; Vanderhagen, Bart; Pelgrims, Bart; Basteyns, An-Valerie; and Delvaux, Freddy R. Characterization of Dark Specialty Malts. *J. Am. Soc. Brewing Chemists*, **2003**, *61*: 125–132. Relationships among malt color and various wort characteristics, including pH.

Coghe, Stefan; Benoot, Koen; Delvaux, Filip; Vanderhagen, Bart; and Delvaux, Freddy R. Ferulic Acid Release and 4-Vinylguaiacol Formation during Brewing and Fermentation. *J. Agric. Food Chem.*, **2004**, *52*: 602–608. Explains how beer gets the phenolic flavor.

Daniels, Ray. *Designing Great Beers*. Brewers Publications, 2000. Good coverage of the beers styles, how they are realized, and what the winning brewers do.

European Brewing Convention. *Analytica-EBC*. 1998. How everything is measured in Europe.

Noonan, Gregory, J. *New Brewing Lager Beer*. Brewers Publications, 2003. A thorough treatment of lager brewing.

Priest, Fergus G.; and Stewart, Graham G. *Handbook of Brewing*, 2nd ed. CRC Press, 2006, Chap. 2. Origin and classification of beer styles.

Richman, Darryl. *Bock*. Brewers Publications, 1994. A detailed discussion of the Bock beer style, its production and history.

QUESTIONS

12.1. Identify and describe the major characteristics that define a beer style.

12.2. How can the brewer control the original gravity?

12.3. How is the color of a malt determined?

12.4. How can the brewer control beer color?

12.5. How do temperature and pressure affect the carbonation level?

12.6. How can the brewer influence the final gravity?

12.7. How can hoppy flavor be introduced?

12.8. What is an ester? What is the usual flavor of an ester?

***12.9.** Beer wort is prepared with 1.5 pounds per gallon Vienna malt. If the brewhouse efficiency is 75%, calculate the specific gravity.

***12.10.** What is the carbohydrate content of wort whose specific gravity is 1.044? How much pale ale malt, at a brewhouse efficiency of 82%, would be needed to make 100 pounds of this wort?

Conditioning vessels. Yuengling Beer Company, Pottsville, Pennsylvania.

CHAPTER 13

FOAM AND HAZE

13.1 SURFACES

There are two beer issues that involve the meeting of gas and liquid or solid and liquid. These are beer foam (gas–liquid) and beer clarity (solid–liquid). It is ironic that these two major concerns in the brewing industry have no effect on the flavor of beer, but only on its appearance. Nonetheless, brewers who fail to satisfy their customers' desire for clear beer with a good head of foam quickly become ex-brewers. Beer foam and clarity became issues only recently in the history of brewing. When beer was ladled out from a barrel to a pottery cup, customers had different expectations.

13.2 SURFACE ENERGY

A **surface** is a boundary where different phases meet, like at the periphery of a bubble. The key concept in dealing with surfaces is **surface energy**. Consider a cube of water one centimeter (0.39 inch) on an edge (about the size of dice). If we were to cut the cube in half with a single slice, we would be creating two square centimeters of new surface. But water is cohesive, its molecules stick together, which is why it stays a liquid and does not fly apart like a gas. To separate the cube of water into two parts, we have to expend energy to overcome the forces holding the water molecules on each side of the slice together. These are the molecules that form the newly created surface (Fig. 13.1).

The Chemistry of Beer: The Science in the Suds, First Edition. Roger Barth.
© 2013 John Wiley & Sons, Inc. Published 2013 by John Wiley & Sons, Inc.

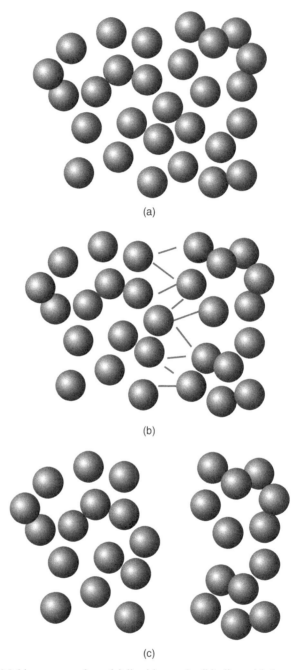

Figure 13.1 Making new surface: (a) liquid sample, (b) slice with forces and, (c) new surfaces.

It takes energy to make the two square centimeters of surface and this energy resides in the surface. If the two half-cubes of water were to come back together, the surface would collapse and the surface energy would be released. Surface energy becomes important when solids or liquids are divided into small particles, or when they are given extra surface by bubbles. The amount of surface energy needed to make a certain amount (like 1 square centimeter) of surface area is called the **surface tension**. The surface energy favors a decrease in surface area. The higher the surface tension, the greater the tendency for particles to grow or combine to reduce the surface area. There are two main ways that this can happen.

> *Disproportionation:* Material from a particle or bubble can dissolve in the liquid, then deposit on a bigger particle or bubble. Many smaller bubbles or particles get smaller until they disappear and a few bigger bubbles or particles get bigger.
>
> *Coalescence:* Two particles collide and become a single particle. In the case of bubbles, the wall between them breaks.

13.3 SURFACTANTS

Surfactants are molecules that migrate to the surface and lower surface energy and surface tension. One end of the surfactant molecule is attracted to one of the phases that meet at the surface and the other end is attracted to the other phase. In foam and haze in beer, one side of the surface is water. The surfactant molecules have a hydrophilic (polar or charged) end that sticks to the water and a hydrophobic end that sticks to the particle or bubble. Molecules with hydrophobic and hydrophilic parts are said to be **amphiphilic**. Figure 13.2 shows a surfactant not normally found in beer called dodecyl sulfate ion, often used as a detergent. The negatively charged sulfate group at the top is the hydrophilic part and the chain of 11 $-CH_2$'s and a $-CH_3$ below is the hydrophobic part. Surfactants tend to arrange themselves so that the hydrophobic part is in contact with the nonpolar bubble or particle, and the hydrophilic part is in the water. The surfactant forms a sort of skin at the surface. This skin stabilizes the particles or bubbles, because when the surface collapses the surfactant can no longer occupy its ideal position, so energy has to be put in.

13.4 HAZE

Haze is cloudiness in beer. Until about 150 years ago all beer was cloudy. Clarity is not an absolute requirement for all beverages. No one expects clear milk or orange juice. Cloudiness is caused by the scattering of light from particles that are larger than the wavelength of light, which is about 0.5

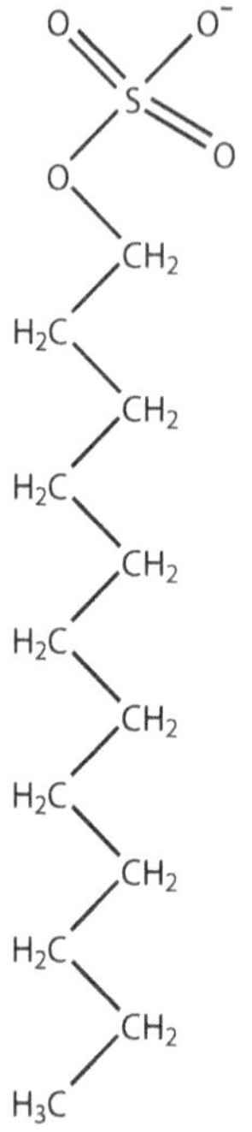

Figure 13.2 Surfactant.

micrometer (20 millionths of an inch). Beer as it leaves the brewery is normally bright and clear. All particles large enough to give haze are filtered out before packaging. The problem arises when very small particles grow by disproportionation or coalescence until they reach the critical size to produce haze. Haze that is present at low temperature and disappears at room temperature is called **chill haze**. This comes about because the increased molecular motion at higher temperature can break the intermolecular forces that hold molecules together to form haze particles. When the temperature goes down, the particles form again. Most haze consists of protein molecules attached to compounds called polyphenols that come from the woody parts of plants. Certain types of proteins are most active for haze formation. Brewers try to minimize haze-inducing proteins and polyphenols by choice of malt, avoiding long contact of barley hulls with hot water, and by treatments with materials, called **finings**, that stick to haze-producing molecules to make particles that can be filtered out. Eventually any beer will become hazy. The problem is worse if the beer is abused by being left in hot places. Brewers put expiration dates on beer to try to keep hazy beer off the market.

Haze can be measured by a technique called light scattering. A beam of light shines into a sample. A light detector is mounted to the side to measure light scattered from the beam. If there are no particles, the light will go in a straight line and none will be scattered to the side detector. But if there are haze particles to bounce the light, some of it will reach the detector, indicating haze. This method is quite reliable, so haze is easy to measure precisely.

Finings

Finings are substances added to the wort or beer to encourage haze-forming materials to settle out. Some finings are added during the boiling process to enhance hot break formation, and some are added during conditioning to help remove haze-forming proteins, polyphenols, and yeast. One important issue with a fining is its isoelectric point (see Section 4.3). The isoelectric point is the pH at which the particles have a net neutral charge. If the solution (wort or beer) in which the finings are used has a pH lower than the isoelectric point, the finings particles will be positively charged; if the pH is higher, they will be negatively charged. Finings work by attracting molecules or particles that we hope to get rid of and forming clumps that fall out or can be filtered out of the beer.

Carrageenan, also called Irish moss, is used during the boil to encourage hot break (Fig. 13.3). It is prepared from certain species of seaweed. A commercial version is Whirlfloc® (Kerry). Carrageenan is a **polymer** consisting of modified units of a sugar called galactose (see illustration opposite Chapter 6 opener). The main modification is the attachment of sulfate groups to certain of the –OH groups on the sugar. In addition, every other sugar has had the –OH groups on carbons 3 and 6 connected to make a second ring. The sulfate group has an isoelectric point of about pH 2, so at ordinary wort pH (pH above 5), carrageenan carries a negative charge. Most proteins in beer have isoelectric points between pH 5 and 6, so they carry positive charges. The positively charged protein molecules are attracted to the negatively charged carrageenan molecules forming nearly neutral particles that clump together and are easily removed with the **trub**.

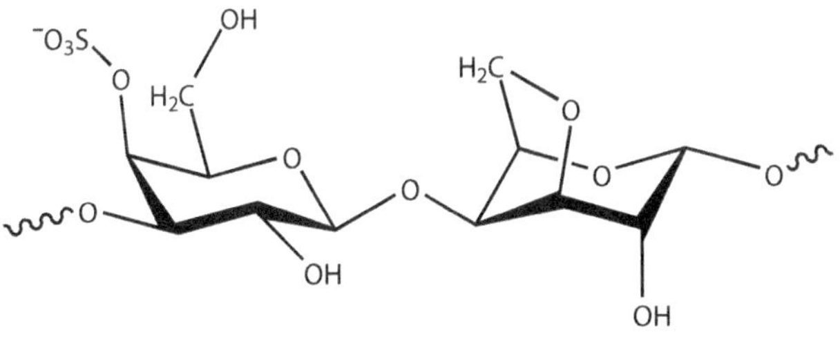

Figure 13.3 Carrageenan.

Collagen is a protein from connective tissue of animals. **Isinglass** is a form of collagen derived from the swim bladders of certain tropical fish. A commercial example is Vicfine® (AB Vickers). **Gelatin** is a form of collagen usually derived from animal skins. These proteins are used during conditioning to encourage the yeast to settle out of the beer. Isinglass is more effective, but gelatin is easier to handle. These proteins have an isoelectric point between 4.8 and 5.5, so in beer, whose pH is about 4, they are positively charged. This allows them to attract yeast cells, which carry negative charges.

Silica gel is a form of highly porous silica (SiO_2) with bound water on its surface. It is made from sand. A commercial product is Britesorb® (AB Vickers). The isoelectric point of silica is 4, which is close to the pH of beer, so silica gel carries little charge. The bound water forms –OH groups on the surface of the SiO_2. The SiO_2 particles are riddled with a network of pores that can admit molecules in a certain size range. The grade of silica gel used for clarifying beer has pores that are suitable for admitting haze-forming protein molecules. Once inside, the molecules stick to the –OH groups by hydrogen bonding. Silica gel works fast; it can be added to the beer as it flows in a pipe to a filtration unit, which removes the silica gel along with the bound protein.

Polyvinylpolypyrrolidone (PVPP) is a plastic with cyclic amide groups attached to a carbon chain with cross-links between chains (Fig. 13.4). A commercial version is Polyclar® (Ashland). PVPP is used in much the same way, and often in conjunction with silica gel. PVPP binds well to polyphenols by offering unshared electron pairs that can accept hydrogen bonds, and by stacking interactions. The polyphenol–PVPP particles are removed by filtration.

Isinglass and gelatin are animal products that can make the beer unsuitable for vegans. In addition, they can be an issue for Jews who follow the kosher laws. Carrageenan, silica gel, and PVPP do not have animal origins.

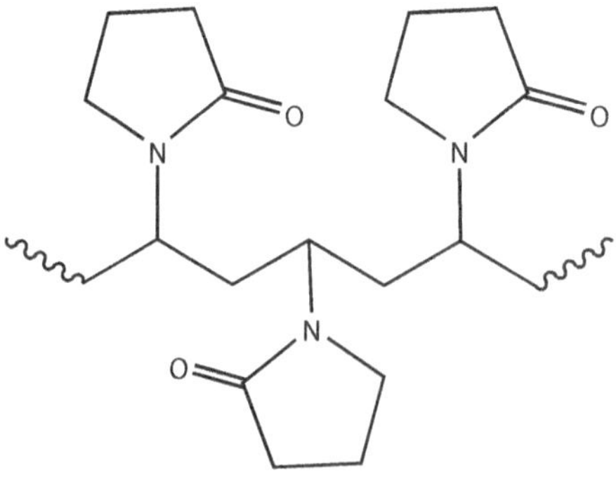

Figure 13.4 Polyvinylpolypyrrolidone (PVPP).

13.5 FOAM

When you open a container of soda and pour it out, bubbles form and briefly make **foam**. The foam soon collapses. By contrast, beer foam lasts for several minutes in a clean glass. To understand what makes beer foam as stable as it is, we need to look at what gives rise to foam in the first place, and what causes it to collapse.

Dissolved Gas

Beer has about 1.8 grams of carbon dioxide dissolved in 12 ounces (355 milliliters) of beer. The amount of a gas (like carbon dioxide) that dissolves in a liquid depends on the pressure of the gas at the surface of the liquid. If we double the pressure, we double the amount that dissolves. This relationship between pressure and amount dissolved is called **Henry's law**. At 40°F (4.5°C) the pressure of carbon dioxide needed to maintain 1.8 grams of dissolved carbon dioxide in a closed 12 ounce can or bottle is 1.7 atmospheres (25 pounds per square inch). When you open the can or bottle, the pressure of carbon dioxide falls to near zero, so all of the dissolved carbon dioxide wants to escape from the beer and go into the gas phase. This comes to a quart of gas (1 liter) coming out of 12 ounces (355 milliliters) of beer. Fortunately, the gas does not all come out at once; otherwise the beer would end up on the ceiling. In order for the gas to come out of solution, many molecules of CO_2 have to come together at the same place and time to give a bubble. Because so much coordinated action on the part of so many molecules is unlikely, it is most common for the gas to come out by making an existing bubble grow, or by forming a new bubble at some site like a scratch or particle that attracts CO_2 molecules. The existing bubble or attractive site is called a **nucleation site**. Pouring out the beer generates little air bubbles that serve as nucleation sites, which is why you usually get foam when you pour beer or soda into a glass.

GASES: A CLOSER LOOK

A gas is a collection of molecules in constant motion. The molecules are far enough apart that each one behaves independently; the forces they exert on one another are very small. The pressure of the gas results from gas molecules striking the walls of the container (or any other surfaces). The pressure exerted by N molecules of gas is N times the pressure exerted by one molecule. As a consequence, the pressure is proportional to the number of molecules, hence the number of moles of gas (see Appendix to Chapter 3, section on "Amount of Substance"). Even if the molecules are of different compounds, the pressure depends only on the total number of them.

(continued)

The force with which the molecules strike the walls and surfaces depends on their kinetic energy, which follows the absolute or kelvin temperature (see Appendix to Chapter 3, section on "SI Units"). As a result, the pressure is proportional to the absolute temperature. The frequency of collisions depends on how far a molecule has to travel to strike a wall. The larger the volume the farther the travel, and the less frequent the collisions. This gives an inverse relationship between pressure and volume. Putting all of this into one equation gives:

$$p = R\frac{nT}{V}$$

where p is the pressure, T is the absolute temperature, V is the volume, and R is a number called the gas constant, whose value depends on the units of measurements and is 0.08206 L·atm·K^{-1}·mol^{-1} when p is in atmospheres (atm), V in liters (L), n in moles (mol), and T in kelvin units (K). The equation is usually stated in this form,

$$pV = nRT$$

called the **ideal gas law**. Because the amount of gas is directly proportional to the volume, amounts of gas are often measured in volume at a temperature of 0°C and a pressure of 1 atmosphere, a condition sometimes called *standard temperature and pressure* (STP). One of the key features of the gas law is that it applies to all gases and all take the same value of the gas constant. By contrast, liquids and solids do not have a universal equation; each substance is different.

Example: Calculate the volume of CO_2 measured at 1.00 atm and 0°C that would be needed to provide 1 kg of beer with 0.5% CO_2.

Solution: The mass of CO_2 is the total mass times the mass fraction: 1000 g × 0.005 = 5 g. The moles of CO_2 is the mass divided by the molar mass (44 g/mol): n = 5 g / 44 g/mol = 0.114 mol. The kelvin temperature is the Celsius temperature plus 273.15 K. For gas law calculations the absolute pressure must be used, not the gauge pressure (pressure in excess of atmospheric).

$$V = \frac{nRT}{p} = \frac{(0.114 \text{ mol})\left(0.08206 \frac{\text{L} \cdot \text{atm}}{\text{K} \cdot \text{mol}}\right)(273 \text{ K})}{1.00 \text{ atm}} = 2.55 \text{ L}$$

Example: Calculate the volume of CO_2, measured at 18°C and 0.970 atm, that can be produced by the fermentation of 100 g of glucose.

Solution: Each mole of glucose yields two moles of CO_2. The moles of glucose is 100 g/180 g/mol = 0.556 mol. The moles of CO_2 is 2 × 0.556 mol = 1.11 mol. The kelvin temperature is 18 + 273 = 291 K. The volume is calculated from

$$V = \frac{nRT}{p} = \frac{(1.11 \text{ mol})\left(0.08206 \frac{\text{L} \cdot \text{atm}}{\text{K} \cdot \text{mol}}\right)(291 \text{ K})}{0.970 \text{ atm}} = 27.3 \text{ L}$$

Gas Law Shortcuts

If some of the variables in the ideal gas law are held constant, simplified expressions result. If a fixed quantity of gas is held at a constant temperature, the right-hand side of the ideal gas law is constant. The ideal gas law reduces to a form called Boyle's law:

$$pV = k \quad \text{or} \quad p_1 V_1 = p_2 V_2$$

Example: A 40 L oxygen cylinder whose initial pressure is 160 atm provides oxygen for wort aeration. What volume of gas will the tank provide at 1 atm?

Solution:

$$V_2 = \frac{p_1 V_1}{p_2} = \frac{(160 \text{ atm})(40 \text{ L})}{(1 \text{ atm})} = 6400 \text{ L}$$

If a fixed quantity of gas is trapped in a constant volume, the ideal gas law can be written

$$p = kT \quad \text{or} \quad \frac{p_1}{T_1} = \frac{p_2}{T_2}$$

Example: A 6000 L empty kettle initially at 1 atm is inadvertently sealed off and allowed to cool from 373 K (100 °C) to 285 K (12 °C). What will be the internal pressure?

Solution:

$$p_2 = \frac{p_1 T_2}{T_1} = \frac{(1 \text{ atm})(285 \text{ K})}{373 \text{ K}} = 0.76 \text{ atm}$$

Gas Mixtures

In an ideal mixture of ideal gases, the mixture acts as though all the molecules were identical. The **partial pressure** of a component of the mixture is the pressure that the component would have if it were alone in the container.

(continued)

Example: Ten grams each of nitrogen (MW = 28 g/mol) and carbon dioxide (MW = 44 g/mol) are loaded into a 5.0 L tank at 6 °C. Calculate the partial pressures.

Solution: We calculate the moles of each gas. For N_2: n = 10 g / 28 g = 0.36 mol. For CO_2: n = 10 g / 44 g/mol = 0.23 mol. The partial pressures are calculated from the ideal gas law.

$$p = R\frac{nT}{V}$$

$$p_{N_2} = \left(0.08206 \frac{\text{L} \cdot \text{atm}}{\text{K} \cdot \text{mol}}\right) \frac{(0.36 \text{ mol})(279 \text{ K})}{(5.0 \text{ L})} = 1.65 \text{ atm}$$

$$p_{CO_2} = \left(0.08206 \frac{\text{L} \cdot \text{atm}}{\text{K} \cdot \text{mol}}\right) \frac{(0.23 \text{ mol})(279 \text{ K})}{(5.0 \text{ L})} = 1.05 \text{ atm}$$

Gases and Liquids

It is important to keep in mind that the gas laws apply only when the gas is all in one phase. If the gas liquefies or dissolves, the ideal gas law and its shortcuts are not even approximately valid.

Vapor Pressure

If we partly fill a container with a liquid then seal it up, some of the liquid will evaporate into the empty space in the container. At the same time, some of the molecules of evaporated liquid (called **vapor**) will collide with the liquid surface and return to the liquid phase. The number of molecules evaporating each second depends on the temperature; at higher temperature more molecules have enough energy to break free of the liquid. Return of vapor phase molecules to the liquid depends on the number of molecules in the vapor space, hence upon the partial pressure. Higher pressure leads to more collisions. The pressure (or partial pressure) of the vapor increases until the rate (speed) of return of vapor molecules to the liquid becomes equal to the rate of evaporation of liquid molecules to the vapor, at which point the pressure stays constant. At any particular temperature, this constant pressure (or partial pressure) is the **vapor pressure** of that substance. The vapor pressure of a liquid depends on the temperature in a complex way.

Dissolved Gases

When a gas comes into contact with a liquid of a different substance, for example, carbon dioxide and water, the gas dissolves to some extent in the liquid. The extent to which the gas can dissolve in the liquid (the solubility) depends on the temperature and the partial pressure of the gas in contact with the liquid. The relationship between partial pressure and solution concentration is called **Henry's law**,

$$c = kp$$

Henry's law states that the concentration of a substance in solution is directly proportional to the partial pressure of that substance in contact with the solution. If we double the partial pressure, we double the concentration. The value of the constant, k, depends on the identities of the gas and the liquid, and on the temperature. The Henry's law constant for carbon dioxide in beer at $10\,°C$ ($50\,°F$) is 0.0798 volumes of gas (measured at standard temperature and pressure — STP) per volume of beer per pound per square inch (psi), 1.17 volumes per atmosphere, 0.0523 mole per liter per atmosphere or 2.30 grams per liter per atmosphere.

Example: Calculate the carbon dioxide pressure in psi needed at $10\,°C$ to carbonate beer at 2.7 volumes CO_2 per volume of beer.

Solution: The pressure is in psi and the solubility is in volumes of CO_2 per volume of beer, so the appropriate value of k is 0.0798 volumes per volume per psi. The solution of Henry's law for (absolute) pressure is

$$p = \frac{c}{k} = \frac{2.7 \text{ v/v}}{0.0798 \frac{\text{v/v}}{\text{psi}}} = 34 \text{ psi}$$

Bubbles

After a bubble forms, it sticks to the nucleation site. All the time that the bubble is submerged in the liquid, it experiences a force that tries to float it to the surface. This force is called **buoyancy**. When the bubble gets big enough, the buoyancy becomes great enough to rip it from the nucleation site and it rises to the liquid surface. At the surface the bubble encounters other bubbles. Bubbles that arrive later push it above the surface. The liquid surrounding the bubble starts to drain down, making the film around the bubble thinner. Unless surfactants can stabilize the foam, it soon collapses (Fig. 13.5). There are many compounds in beer that enhance foam stability. Some of these are proteins and some are compounds from the hops. Most proteins have potential

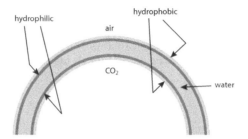

Figure 13.5 Surfactant on bubble. (See color insert.)

amphiphilic properties and can act as surfactants, but they are usually folded with the hydrophilic parts on the outside (in the water) and the hydrophobic parts on the inside, in contact with one another. Boiling unfolds the proteins and may make them better foam stabilizers. When more than one surfactant is present, they can reinforce one another or they can have a negative effect. Some surfactants stick to one another better than they stick to a bubble surface, so surfactants that may be effective foam stabilizers one at a time are ineffective together. This is why traces of detergent in beer glasses can kill the foam. Greasy or oily substances, which could come from incomplete washing of serving glasses or from the mouths of the drinkers who are also munching out on wings or fries, are very effective foam killers. The hydrophobic parts of the surfactants from the beer stick to the oil better than they do to the gas layer in the bubbles. This takes the surfactant away from the bubble surface, so the foam collapses. It doesn't take much.

Stability of beer foam is called **head retention** in brewing jargon. There are many methods to measure head retention. This is a sure sign that there is no really satisfactory method. Some methods involve measuring the distance that the head of foam goes down in a certain time, some involve measuring the amount of beer that drains out of a certain amount of foam in a certain amount of time. The best that can be said is that most of the methods place the same beers in the same order in terms of head retention. Too much head retention can be worse than too little. The customer wants a glass of beer with an attractive layer of foam, not a glass of foam with a little beer at the bottom. Another issue is that the attractiveness of the head of foam is not simply an issue of its volume. The color, bubble size, and other factors play a role. Making long-lasting foam that the customer finds uncool for some reason does not solve the head retention problem.

13.6 FOAM ISSUES

Foam is enhanced by hydrophobic proteins or fragments. Wheat malt seems to be a rich source of these; a small amount of wheat in the grist can give a marked increase in foam. Hop bitter compounds enhance foam. Acidity (low pH) is good for foam, possibly because at low pH the acidic hop compounds are less ionized, hence more hydrophobic. The presence of nitrogen gas enhances foam by suppressing the migration of gas out of small bubbles. The use of filtration rather than lautering for wort separation has been found to enhance foam.

Fats, detergents, and long-chain alcohols decrease foam. Ethanol enhances foam at low levels, but decreases it at the normal levels found in beer. Certain grains, like oats, tend to raise the level of fats in the beer, lowering the foam. High gravity brewing, which is brewing with concentrated wort followed by dilution before packaging, decreases foam.

Gushing

Gushing is the escape of nearly all of the carbon dioxide from beer at once. This can lead to a good part of the beer spurting out when the container is opened. There are many possible causes of gushing, including too much carbonation, opening at high temperature, shaking, and the presence of bubble nucleation sites. Nucleation sites can be produced by bacterial contamination, by glass dust from new bottles, or by proteins in the malt deriving from a fungus-induced barley disease called *Fusarium* head blight. Maltsters will not knowingly accept such barley. Whatever the cause, gushing is a fatal defect; the beer is undrinkable. Gushing can be particularly exasperating because it shows up intermittently. One batch gushes, the next is fine. It seems to have outbreaks, then goes into remission.

13.7 NITROGEN AND WIDGETS

Certain styles of beer, most notably British ales, are packaged with low levels of carbon dioxide. Because carbon dioxide is acidic, it lends a characteristic flavor to the beer. Nitrogen, which is neutral and has no flavor, is sometimes mixed with carbon dioxide as a foam-raising gas in beer. This gives an enhancement in the quality of the foam, especially for beer styles packaged with low levels of carbon dioxide. Nitrogen is about 100 times less soluble in water than carbon dioxide. The disproportionation process, which is the major mechanism for foam loss, depends on gas in small bubbles dissolving. Dissolving of nitrogen is very limited, so it barely participates in disproportionation. The carbon dioxide tends to migrate from smaller to larger bubbles, but the small bubbles don't completely disappear because the nitrogen in them can't escape into solution.

Nitrogen allows the gas pressure to be increased without increasing the volume of gas dissolved in the beer. The extra pressure can be used to drive the beer through narrow openings to produce small bubbles that serve as nucleation sites. This is the principle of the **widget**, a hollow plastic container with a small hole. Air inside the widget is replaced with nitrogen and the widget is put into the beer can or bottle. The can is filled and nitrogen is added under pressure, driving beer into the widget and compressing the gas that is already in the cavity of the widget. When the package is opened, the compressed gas drives the beer out through the opening, producing a creamy head of foam.

BIBLIOGRAPHY

Bamforth, C. W. The Relative Significance of Physics and Chemistry for Beer Foam Excellence. *J. Inst. Brew.*, **2004**, *110*(4): 259–266. A readable (albeit with several equations that the reader can ignore and take Bamforth's word for it) article about the effect of various factors on foam.

Briggs, Dennis E.; Boulton, Chris A.; Brookes, Peter A.; and Stevens, Roger. *Brewing Science and Practice*. CRC Press, 2004, pp. 697–711. All about haze.

Hough, J. S.; Briggs, D. E.; Stevens, R.; and Young, T. W. *Malting and Brewing Science*, 2nd ed. Aspen, 1999, Volume 2, pp. 811–832. All about foam.

Rehmanji, Mustafa; Gopal, Chandra; and Mola, Andrew. Beer Stabilization Technology. *MBAA Tech Quarterly*, **2005**, *42*(4): 332–338. Haze and controlling it with finings.

QUESTIONS

13.1. What is surface energy?

13.2. What is a surfactant?

13.3. Identify and give brief explanations for the two mechanisms by which surface area is lost.

13.4. Why do greasy substances kill beer foam?

13.5. What is Henry's law?

13.6. What is head retention?

13.7. What is haze?

13.8. What are isinglass, gelatin, silica gel, and PVPP? How is each used?

13.9. What are some causes of gushing?

13.10. When a can of beer is shaken just before it is opened, the beer gushes. What nucleation sites does the shaking produce?

13.11. What is a widget and how does it work?

*13.12. Given that 150 grams of glucose is allowed to ferment in 21 liters of beer, if all the CO_2 dissolves, what will be the carbonation level in grams of CO_2 per liter?

*13.13. A 5000 liter batch of beer wort is to be oxygenated at a rate of 0.01 gram per liter. The molar mass of oxygen is 32 g/mol. Calculate the volume of oxygen required measured at 7°C (280 K) and 1.00 atmosphere.

Stacked kegs. Yuengling Beer Company, Pottsville, Pennsylvania.

CHAPTER 14

BEER PACKAGING

The brewer's job is not over until the beer is put into a suitable package that allows it to be transported and stored with minimal deterioration. Modern beer has to be kept under pressure, otherwise it loses its carbon dioxide and goes flat. Beer is shipped in barrels called **kegs** and **casks**, or serving size containers known collectively as **small packs**. Small packs can be glass bottles or aluminum cans. Efforts are being made to switch from glass to plastic bottles, but so far they have not been widely adopted. The major technical problem for plastic is that small amounts of oxygen can flow through it. This can lower the shelf life of the beer to an unacceptable level. Also, plastic bottles are not well adapted to heat treatment in a pasteurizer. Many consumers think plastic is uncool, so even if all technical problems were solved, it could take time for plastic to achieve market acceptance. Sports arenas, where aluminum cans could be used as projectiles, may become an attractive market for plastic beer containers.

14.1 CASKS AND KEGS

Casks are descendants of the wooden barrels that were once used to ship beer. Wooden casks are seldom used today except for promotion and ceremonial occasions because of their weight and the difficulty and expense of keeping them clean and in good condition. Today, metal casks made either of aluminum or of stainless steel are used. The cask is essentially a barrel, usually made of metal, with a hole on the end (the **keystone** hole) and a hole in the side (the

---

The Chemistry of Beer: The Science in the Suds, First Edition. Roger Barth.
© 2013 John Wiley & Sons, Inc. Published 2013 by John Wiley & Sons, Inc.

bung hole). Some casks have additional openings so they can be conveniently tapped in alternate positions. The cask serves both as a shipping container and as a conditioning vessel. Beer with suspended yeast is put into the cask through the bung hole. Sugar may be added to give carbonation. **Finings** (see "Finings" in Section 13.4) are added to help the yeast settle out. The cask is stoppered with a **bung** and shipped to a pub, which is where the final phase of conditioning takes place. The handling of casks at the pub requires skill and constant attention. When the cask is set up, a **tap** (valve) is hammered into the keystone hole and a peg called the **spile** is driven into the bung. The spile is adjusted to give a gas leak rate that keeps the correct pressure. Brewers tend to be reluctant to depend on the skill of outsiders to maintain the quality of their product. For a while there was a trend to do away with casks. This met with consumer resistance in the form of an organization called the **Campaign for Real Ale**, or **CAMRA**. British brewers continue to provide ale in casks so pubs can serve ale with a real cask-conditioned flavor. The brewers have found it to be in their interests to maintain good relations with CAMRA.

The alternative to the cask is the **keg** (Fig. 14.1). Kegs are cylindrical vessels made of metal, usually stainless steel, although aluminum can be used if it is coated with plastic on the inside. The usual size of a keg in the United States is 15.5 U.S. gallons (58.7 L), which is half of a beer barrel. The usual European keg is 50 liters. Smaller sizes are used for parties. A keg has a single opening on the end. The opening leads to a tube called the **spear** that goes to the bottom of the keg. At the top of the spear is a valve that keeps the keg closed when it is not connected to dispensing gear. The spear also has an opening near the top to let gas in to drive the beer out. Putting the keg into service involves installing a valve called a **keg coupling** into the keg opening so that a connection for beer is made to the bottom of the spear and a connection for

Figure 14.1 Keg washer. Victory Brewing Company, Downingtown, Pennsylvania.

gas, usually carbon dioxide, is made to the hole near the top of the spear. Beer sold in kegs is fully conditioned. It may be pasteurized or sterile-filtered on the way to the keg to remove any microbes. If the beer is intended for local use, it may be kegged without sterilization, in which case it must be kept cold.

14.2 GLASS

Glass bottles have been around for about 3600 years. Glass is highly resistant to corrosion or chemical attack. It is made from sand, which is silicon dioxide (SiO_2). Glass made from pure silicon dioxide is called **fused quartz**. Fused quartz is used in very high temperature applications like projector bulbs because it melts above 2300 °C (4200 °F). This makes it too difficult to work with for beer bottles. Regular glass is made by mixing the sand with a **flux**, a substance that lowers the melting point. Almost anything will lower the melting point of anything it dissolves in, a phenomenon called **freezing point depression**. For example, salt will lower the melting point of water, which is why streets are salted in the winter. The usual flux for common glass is sodium carbonate (Na_2CO_3) plus lime (calcium oxide, CaO) or limestone (calcium carbonate, $CaCO_3$). These lower the melting point to a manageable 1500 °C (2730 °F), which can be attained in a gas flame. This type of glass is called **soda lime glass**. The sodium carbonate without the lime would do the job as a flux, but the resulting glass would be water soluble. Beer bottles are usually made of brown glass to protect the beer from light. Compounds containing carbon or nickel are added to make the glass brown.

Glass is formed into bottles by blowing. A gob of melted glass is blown with air into a bubble in a mold that defines the outer contour of the bottle. The neck of the bottle has a ridge to give the cap something to grip. For twist-off caps, the ridge goes in a spiral. Because the spiral is not very thick, twist off bottles are not reused. Decorations and advertisements to promote the beer brand usually are applied as labels because glass is not easy to decorate. Many bottles have labels on the front, back, neck, and sometimes shiny foil wraps around the cap. None of this has any effect on beer quality, but it does help identify the brand to the consumer. It is said that packaging is everything. With a standardized commodity like commercial beer, no brewer wants to miss an opportunity to give the product some class.

Glass consists of chains and rings formed from alternating silicon and oxygen atoms: –O–Si–O–Si–O–. The chains and rings are randomly distributed giving what is called an **amorphous** (Greek: *a-* without, *morphe* form) structure. The atoms in glass are fixed in place by strong covalent bonds. When glass is cooled from the high temperatures at which it is formed, it gets more and more viscous (slow flowing), but there is no clear-cut freezing point. Eventually it becomes stiff enough to support its own weight and begins to behave like a solid. During cooling, the glass shrinks. Microscopic cracks develop at the surface. When enough tension (stretching force) is applied, these cracks can grow rapidly and the material shatters, sometimes with impressive violence. The tendency to crack under tension is called **brittleness**; glass is brittle. Glass beer bottles need to be made thick enough to minimize the risk of shattering because the gas pressure keeps them under tension.

14.3 METALS

Metals are elements with loosely held valence electrons that tend to break away, leaving a positive ion. Most metals occur in nature in the form of ionic compounds. Iron is dug up in the form of compounds containing Fe^{2+} or Fe^{3+} ions, like Fe_2O_3, known to miners as hematite. The most useful ore of aluminum is Al_2O_3, called alumina. You can't package beer in a bunch of ions; first the electrons have to be driven back onto the ions to give the neutral metals. The chemical reaction in which an electron is added to something to give a more negative or less positive product is called reduction. The equations for the reduction of copper(II) ion to copper metal is $Cu^{2+} + 2e^- \rightarrow Cu$. Electrons are subatomic particles; they don't come in sacks or bottles. Any electrons that go into one reactant must come from another. The reduction of copper must be coupled with a reaction in which electrons are released, a process called oxidation. The equation for the oxidation of zinc is $Zn \rightarrow Zn^{2+} + 2e^-$. The oxidation and reduction reactions are called half-reactions because a reduction cannot occur without an oxidation. The full equation for the reduction of copper(II) ion by zinc is $Cu^{2+} + Zn \rightarrow Cu + Zn^{2+}$. This type of reaction is called an oxidation–reduction, or **redox**, process.

Most metals can be reduced (gain electrons) by treatment, usually at high temperature, with a substance such as carbon that can give up electrons. This works for metals that can accept electrons fairly easily, like iron or copper. Some metals, like aluminum, accept electrons only with great difficulty, so special methods are needed. Once a metal is reduced, it has a tendency to give up electrons and become oxidized. Because we live in an atmosphere with 20% oxygen, there is always something to accept the electrons that the metal gives up: $O_2 + 4e^- \rightarrow 2O^{2-}$. Oxidation of metal results in loss of material and the formation of holes, a process called **corrosion**. Corrosion can be prevented by applying a coating to the metal to keep away anything that can accept electrons. Metal small-pack containers are often coated with plastic on the inside to prevent corrosion. Another way to prevent corrosion is to mix the metal with another metal that can oxidize, but the resulting compound sticks to the metal to form a coating. This is how stainless steel works. The type of stainless steel usually used in beer kegs has about 20% chromium and 10% nickel. When oxygen is present, these metals form oxides that migrate to the surface and form a tight coating that protects the iron underneath.

The loosely held valence electrons in metals give rise to a special type of bonding called metallic bonding. The valence electrons in hunks of metal form a sort of cloud that can move nearly freely throughout a piece of metal. We can think of a metal as being composed of an array of positively charged atom cores (nucleus plus core electrons) interspersed with freely moving valence

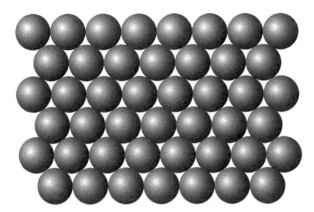

Figure 14.2 Crystalline structure.

electrons. These freely moving electrons give rise to the familiar metallic properties. When an electrical potential (voltage) is applied to the metal, the electrons move to the positive terminal, giving an electrical current. This accounts for the electrical conductivity of metals. If one end of the metal is at a higher temperature than the other, the electrons at the hot end gain energy and, because they move freely, carry it to the lower temperature end. This accounts for the thermal conductivity of metals. When light tries to enter a metal, the freely moving electrons cancel the rapidly changing electric field that constitutes light. This prevents the light from entering the metal. Instead the light is reflected by the metal, which accounts for metallic luster. When a piece of metal is subjected to a force that tends to move the atoms out of their lowest energy positions, the electron cloud can redistribute the charge keeping the energy in bounds. This gives metals **toughness**, that is, resistance to cracking and shattering. Toughness is the opposite of brittleness.

Metals nearly all have an orderly three-dimensional arrangement of atoms called a **crystalline structure** (Fig. 14.2). In actual metals, the arrangement is not perfect; there are gaps in the crystal called **defects**. The defects allow the atoms to move in response to a force. The crystalline arrangement thus allows metals to stretch, compress, and bend without shattering in response to forces. This gives metals desirable mechanical properties, particularly the ability to bend and stretch without tearing or cracking. The crystalline arrangement of the atoms coupled with the freely moving valence electrons allow metal cans to be strong under the tension produced by the carbonation pressure. In fact, this tension is necessary to keep the can stiff enough so that it is not crushed by cans sitting on top of it. An empty can has very little strength; it resembles aluminum foil.

14.4 ALUMINUM

Aluminum is a much more recent material than glass. When the Washington Monument was capped with aluminum in 1884, one ounce of aluminum cost as much as a person's daily wage. Until improvements in producing aluminum brought the price down, it was not practical for use in routine packaging. Aluminum exists in nature in the form of Al^{3+} ions. It is difficult to get these ions to accept electrons and go to the metal. The reaction that needs to be done is $Al^{3+} + 3e^- \rightarrow Al$.

You can't buy electrons; any electrons that are used by a reactant in one part of a reaction have to be provided by a reactant in another part. Until about 1888, the electrons were provided by sodium metal. Sodium is in group IA of the periodic table; it gives up its one valence electron easily. The complete reaction for the production of aluminum was $AlCl_3 + 3Na \rightarrow Al + 3NaCl$. This process requires about 2½ pounds of sodium metal for each pound of aluminum produced. Sodium is difficult to isolate as an element and it is dangerously reactive. In 1888 a new method, called the Hall–Heroult process, was introduced in which electrons derived from carbon are driven by an applied voltage. This is how aluminum is made today. The technological blockade that the Hall–Heroult process overcame was getting the aluminum ore into solution. The voltage needed to reduce aluminum would rip water apart to oxygen and hydrogen. Aluminum comes in the form of aluminum oxide (Al_2O_3), which melts above 2000°C (3600°F). In the Hall–Heroult process the aluminum oxide is dissolved in a melted ionic compound called cryolite (Na_3AlF_6). Cryolite melts at 1000°C (1800°F) and is a good solvent for aluminum oxide. Aluminum fluoride (AlF_3) is added to lower the melting point by freezing point depression. The electrical current is introduced through hunks of carbon, which is a fair conductor of electricity. The aluminum formed is liquid; it sinks to the bottom of the container. The carbon electrode reacts with the oxygen from the aluminum oxide to give carbon dioxide, which is vented to the atmosphere.

It takes about 13 kilowatt hours (48 million joules) of electrical energy to make one kilogram (2.2 pounds) of aluminum from pure aluminum oxide. At an average U.S. price for electricity of 10¢ per kilowatt hour, the electricity cost is $1.30 per kilogram. The price of aluminum is about $2 per kilogram, so the electricity dominates the cost. Aluminum can be recycled with a much smaller energy input.

Aluminum used for cans contains small amounts of magnesium and manganese to improve its mechanical properties. The cans start out as a huge roll of sheet aluminum, which is punched to give shallow cups. Each cup is driven through progressively smaller holes, getting taller and thinner each time. The final thickness of the can walls is 0.0035 inch (0.09 millimeter). The bottom of the can is pressed inward into a dome to help it resist the pressure of the beer. All of this takes about a 25th of a second (40 milliseconds). The top of the can can't be domed because the pull-tab has to be on a flat surface. Flat surfaces don't withstand pressure well, so the top of the can has to be thicker than the rest. To minimize the use of thick metal, the upper part of the can is pressed in to narrow the opening to accommodate a small top. The can is lined with a thin coat of plastic to prevent the beer from reacting with the bare metal. The

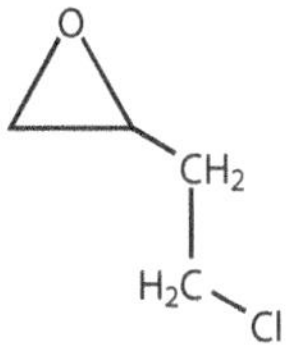

Figure 14.3 Epichlorohydrin (epoxide).

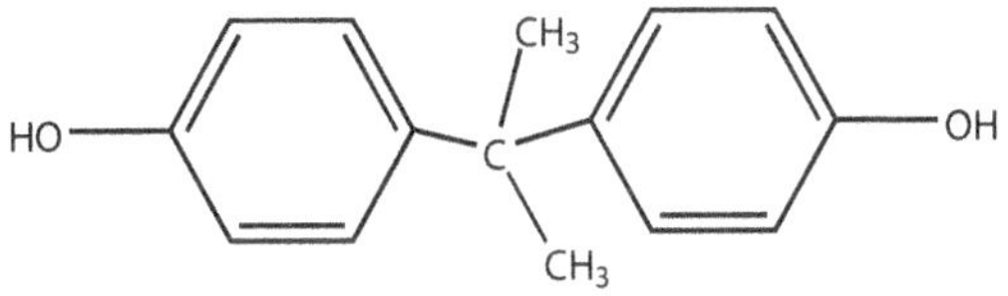

Figure 14.4 Bisphenol A.

top of the can is installed after filling. Before the can is shipped to the brewer, it is decorated with characteristic designs. This is convenient because it saves the brewer the trouble and expense of applying labels after the filling process.

A concern has been raised about the coating that is applied to keep the beer (or other product) from making direct contact with the metal. The coating is a type of polymer called an epoxy. An epoxy is a plastic formed from the reaction of molecules having an epoxide group (Fig. 14.3) with molecules having an amine functionality. These form tough, unreactive materials that are used for coatings of food containers. The concern is that one of the compounds used to make the epoxy polymer is bisphenol A (BPA, Fig. 14.4). The other is epichlorohydrin. Bisphenol A has been implicated in a number of human diseases. A tiny, but detectable amount of it leaches into whatever is in a container coated with epoxy. The level of bisphenol A in beverage cans is well below regulatory standards, but the concern exists nonetheless. Efforts are under way to commercialize alternatives to coatings containing bisphenol A.

Cans are inexpensive and very light, are not susceptible to breaking, and keep light out. They can be taken to places where glass would not be welcome, like the beach or a swimming pool. Cans stack well, saving space and energy in shipping. A disadvantage of cans is that during pouring, the beer makes contact with the top of the can, which may or may not be clean. Many beer drinkers prefer beer from bottles. Bottles have a classier image, and with beer, image can be the deciding factor.

14.5 BOTTLING AND CANNING

The key issues in filling small pack containers are that the process must be very fast and must not allow any air to enter the beer. Bottle fillers work at 1000–1600 bottles a minute; can fillers work at 2000 cans a minute (Fig. 14.5).

Figure 14.5 Packaging hall. Victory Brewing Company, Downingtown, Pennsylvania.

Figure 14.6 Crown caps.

Before a container is filled with beer, it is first filled with carbon dioxide so the inside pressure is equal to the beer pressure. Transferring the beer without a pressure change reduces foaming during the filling process. When bottles are filled, the air is pumped out, and then carbon dioxide is pumped in under pressure, then the beer is put in. Cans can't be pumped out or they would collapse, so the air is pushed out with carbon dioxide. Immediately after filling, the package is closed. In the case of bottles, the closure is a **crown cap** (Fig 14.6). Crown caps start out as slightly cupped disks with 21 folds around the edge. A plastic washer inside the cap makes a gas-tight seal. The cap is placed on the bottle and the edges are pushed down until they grip the ridge on the rim of the bottle. In the case of cans, the lid is positioned on the can, then a clever arrangement of rollers rolls the edge of the can around the edge of the lid. A second set of rollers flattens the crease to make a pressure-tight seal.

14.6 MICROBE REDUCTION

There are three approaches to suppression of bacteria and fungus in beer. Hop bitter compounds, acidity, and alcohol provide beer with significant resistance to spoilage by microbes. Beer shipped in casks is not treated at all, relying on beer's inherent stability. Some keg beer and bottled beer from craft breweries is also untreated. The shelf life of these beers is shortened, but customers are willing to accept this limitation in the interest of drinking fresh beer. Mainstream commercial breweries either run the beer through very fine filters that remove microbes or heat the beer to about 60 °C (140 °F) to kill most microbes, a process called pasteurization. Filtering must be done before filling; pasteurization can be done either before or after filling. If the microbes are eliminated before the beer is packaged, the packaging must be done under sterile conditions. All microbe reduction processes have some effect on the flavor of the beer.

BIBLIOGRAPHY

Goldhammer, Ted. *The Brewer's Handbook*, 2nd ed. Apex, 2008, Chaps. 16–18. Carbonation, bottling, and kegging.

Lewis, Michael J.; and Young, Tom W. *Brewing*, 2nd ed. Springer, 2001, Chap. 20. Packaging and dispensing beer.

QUESTIONS

14.1. Explain the difference between a cask and a keg.

14.2. What is the function of finings?

14.3. What are the raw materials for glass?

14.4. What are some advantages and disadvantages of aluminum cans compared with glass bottles?

14.5. What is the major contributor to the cost of aluminum?

14.6. What is freezing point depression?

14.7. Describe three ways that packaging can help to protect or prolong beer quality.

14.8. How is aluminum ore made into metallic aluminum?

14.9. Define brittleness and toughness. Identify some physical properties characteristic of metals. Explain how these derive from the electronic structure of the metals.

Bottles, caps not yet crimped.

CHAPTER 15

BEER FLAVOR STABILITY

Over time, the properties of beer change, mostly for the worse. Ultimately, the beer becomes stale. Brewers are very interested in maintaining high quality in their product for as long as possible. In an effort to keep stale beer off the market, brewers mark every beer container with a "best by" date and a lot number. Despite much research, there is little agreement on most of the basics of flavor stability. The characteristics of staleness depend on the beer style and on the details of its production and storage. We will touch upon a few aspects of flavor stability.

15.1 TYPICAL FLAVOR CHANGES

After packaging, beer begins to lose bitterness and gain sweet and toffee-like flavors. These changes keep increasing. At a certain point, there is a rapid growth in a flavor officially called **ribes** (RYE bees), after *Ribes nigrum*, the black currant. Most beer drinkers refer to ribes as catty, or cat pee. The catty flavor reaches a maximum, then begins to decline, eventually disappearing. Around when the catty flavor reaches its maximum, a cardboard or straw-like flavor slowly starts to become noticeable. The cardboard flavor increases continuously. In pale lager beer the cardboard flavor becomes the dominant indication of staleness. Bitter aftertastes, metallic flavor, burnt flavor, and other flavors are reported in some styles under some storage conditions.

The Chemistry of Beer: The Science in the Suds, First Edition. Roger Barth.
© 2013 John Wiley & Sons, Inc. Published 2013 by John Wiley & Sons, Inc.

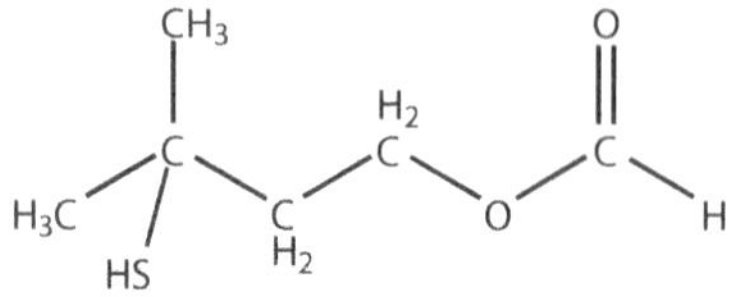

Figure 15.1 (3-Mercapto-3-methylbutyl) formate.

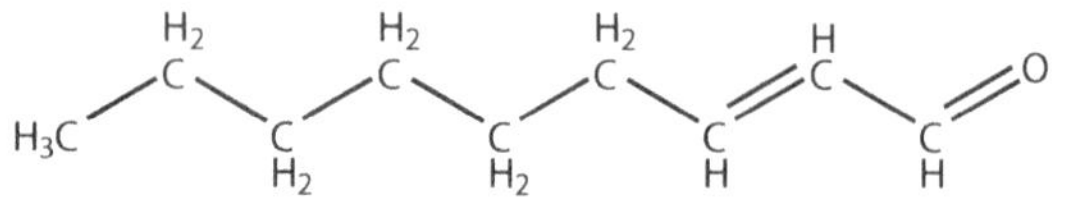

Figure 15.2 Trans-2-nonenal.

The catty flavor has been attributed to several compounds, including (3-mercapto-3-methylbutyl) formate (Fig. 15.1). Little is known about how it arises in beer and how it is removed. Its formation depends on the presence of oxygen in the packaged beer.

The best known staling compound in beer is trans-2-nonenal, officially (E)-2-nonenal (Fig. 15.2). This aldehyde gives a cardboard flavor at very low concentration. Other compounds with the carbonyl group (C=O), especially those that also have a C=C double bond, are also likely to contribute stale flavors. Trans-2-nonenal, if present during fermentation, accepts hydrogen atoms to give a nearly tasteless alcohol. Oxygen in packaged beer does not affect the production of trans-2-nonenal. It is possible that the compound is formed earlier in brewing, gaining protection from the fermentation process by binding to amino acids, and is then released during storage of the beer.

15.2 THE ROLE OF OXYGEN

The one thing about flavor stability that everyone agrees on is that oxygen is bad for it. If oxygen is rigorously excluded from the packaged beer, the catty flavor does not arise at all. Excluding oxygen from earlier stages, like milling and mashing, may help to reduce the cardboard flavor. To get an idea of what is going on, we need to consider the structure and chemistry of oxygen.

Oxidation and Reduction

Reactions involving the transfer of electrons are called oxidation–reduction, or redox, reactions. The species (atom, molecule, ion, or radical) that gives up electrons is said to be oxidized, or to undergo **oxidation**. If something gives up electrons, some other species has to accept them. The species that accepts the electrons is said to be reduced, or to undergo **reduction**. Oxygen has a tendency to accept electrons, that is, it becomes reduced. Its reaction partner becomes oxidized. Consider the reaction $4Na + O_2 \rightarrow 4Na^+ + 2O^{2-}$. The oxygen on the left is neutral, it has no charge. During the reaction it picks up four

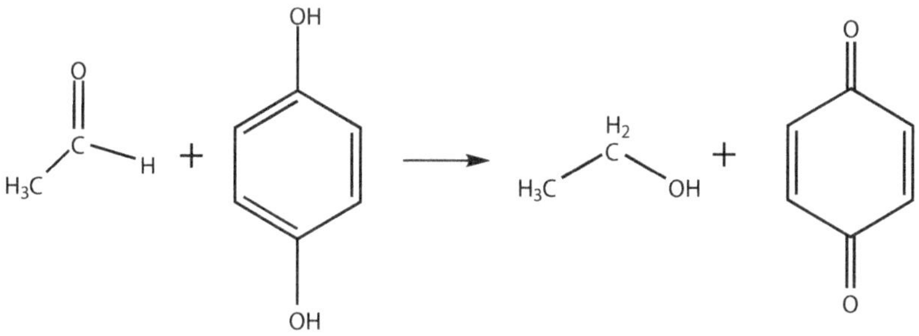

Figure 15.3 Oxidation–reduction: acetaldehyde + hydroquinone → ethanol + benzoquinone.

electrons, yielding the two oxide ions on the right, each of which has a –2 charge. Each oxygen atom has accepted two electrons for a total of four electrons. Each sodium atom has provided one electron for a total of four electrons. It is easy to see where the electrons are coming from and going to.

Now consider the reaction $2H_2 + O_2 \rightarrow 2H_2O$. It turns out that this is also considered an oxidation–reduction reaction. A reaction involving the addition of H atoms is a reduction because we count the electrons that come with the hydrogen atoms. The removal of H atoms is an oxidation. Transfer of the hydrogen ion, H^+, is neither oxidation nor reduction; it is an acid–base reaction. In the reaction shown in Figure 15.3, acetaldehyde gains two hydrogen atoms yielding ethanol. Acetaldehyde is reduced; hydroquinone is oxidized.

OXIDATION–REDUCTION: A CLOSER LOOK

Oxidation is defined as the loss of electrons; reduction is the gain of electrons. In a covalent bond, electrons are shared. To keep track of electrons we need a scheme to allocate the electrons among the atoms that share them. The scheme used by chemists is **oxidation number** (ON). An oxidation number is a fictitious charge on each atom that we use to keep track of electrons. To determine the oxidation number we subtract the number of allocated electrons from the periodic table group number of the atom. Each atom is allocated all its unshared electrons. Shared electrons are allocated to the more electronegative (see Table 3.4) of the atoms sharing them. Electrons shared by atoms of the same element are divided equally between them. An increase in oxidation number shows that the atom has been oxidized; a decrease in ON shows that it has been reduced. The sum of the ONs on a molecule or ion is equal to the charge. The ON of a solitary ion is equal to its charge. The atoms in a neutral element (like O_2) have an ON of zero. For complicated cases, draw the Lewis structure showing all valence electrons. For elements common in beer, the order of electronegativity is $O > N > S > C > H > P$. Circle each

(*continued*)

atom in a way that puts all unshared electrons inside the circle and puts shared electrons in the circle of the more electronegative atom.

Example:

$$2Fe^{3+} + CH_3CH_2OH \text{ [ethanol]} \rightarrow CH_3CHO \text{ [acetaldehyde]} + 2Fe^{2+} + 2H^+$$

The ONs of the one-atom ions (Fe^{3+}, Fe^{2+}, and H^+) are equal to the charges on the ions. Ethanol and acetaldehyde are complicated cases, so we will work out their ONs step by step. We first draw the Lewis structure of ethanol with all valence electrons shown as dots (Fig. 15.4). Oxygen is the most electronegative atom, so we circle it and include its four unshared electrons and four shared electrons within its circle. Carbon is next in electronegativity. The electrons shared with hydrogen go into the circle with the carbon. The two electrons shared between the carbon atoms are divided equally, one to one carbon atom and the other to the other. Hydrogen is the least electronegative element in this compound, so the hydrogen atoms get none of the shared electrons. The result from Figure 15.4 for ethanol is that we allocate 7 electrons to the methyl (CH_3) carbon, so its oxidation number is $4 - 7 = -3$. The carbon with the –OH is allocated 5 electrons for an ON of $4 - 5 = -1$. The oxygen is allocated 8 electrons for an oxidation number of $6 - 8 = -2$. The hydrogens are allocated zero electrons for an ON of $1 - 0 = +1$.

In acetaldehyde the same analysis (Fig. 15.5) gives an ON of $4 - 7 = -3$ for the methyl carbon, $4 - 3 = +1$ for the carbonyl carbon (C=O), and $1 - 0 = 1$ for the hydrogen atoms.

Looking at the original reaction equation, we compare the reactants to the products. During the course of the reaction, the iron ion went from an ON of +3 to +2; it is reduced. The hydrogen atoms in ethanol, acetaldehyde, and the H^+ ions all have ONs of +1, so they are neither oxidized nor reduced. The methyl carbon on ethanol and that on acetaldehyde both have an ON of –3. The oxygen atoms on ethanol and acetaldehyde both have an oxidation number of –2. The carbon attached to the oxygen atom in ethanol has an oxidation number of –1, compared to that in acetaldehyde, which has an ON of +1. This carbon atom is oxidized from

Figure 15.4 Allocating electrons: acetaldehyde.

Figure 15.5 Allocating electrons: ethanol.

TABLE 15.1

	Fe	C_m	C_O
Reactants	+3	−3	−1
Products	+2	−3	+1
Change	−1	0	+2
Number	2	1	1
Total change	−2	0	+2

an ON of −1 to an ON of +1. It lost two electrons during the course of the reaction. These electrons were transferred to the two Fe^{3+} ions, reducing them to Fe^{2+}. Table 15.1 summarizes the changes in oxidation numbers on the iron ion, the methyl carbon (C_m), and the oxygen-bearing carbon (C_O).

The last row shows that the total gain in oxidation number on the part of C_O is equal to the loss on the part of the two iron ions (Fe).

Oxygen Structure

Each oxygen atom has six valence electrons. The twelve valence electrons in the O_2 molecule form two shared pairs (a double bond) and four unshared pairs (Fig. 15.6). If we allocate the shared electrons equally to the two atoms, we can assign six valence electrons to each oxygen atom in the oxygen molecule.

Oxygen can gain electrons to form the oxide ion, O^{2-}. Because it is strongly basic, the oxide ion usually grabs a hydrogen ion from water to form the hydroxide ion: $O^{2-} + H_2O \rightarrow 2OH^-$. The oxygen atoms in the oxide ion and the hydroxide ion have eight valence electrons, compared to six for free oxygen. Each oxygen atom gained two electrons; the two atoms in the oxygen molecule gained a total of four electrons. In addition, two bonds broke. A transformation this complicated does not happen all at once.

Reactive Oxygen Species

The path from the oxygen molecule to oxygen-containing organic compounds like aldehydes and ketones involves a number of intermediates called reactive oxygen species, or ROS. Most of these are **free radicals**, that is, they have unpaired electrons. Because free radicals can't have filled or empty valence shells, they are fundamentally unstable. The natural tendency of electrons to form pairs also makes it easy for free radicals to react. Organic free radicals are most stable when the unpaired electron forms on a carbon one removed

Figure 15.6 Oxygen.

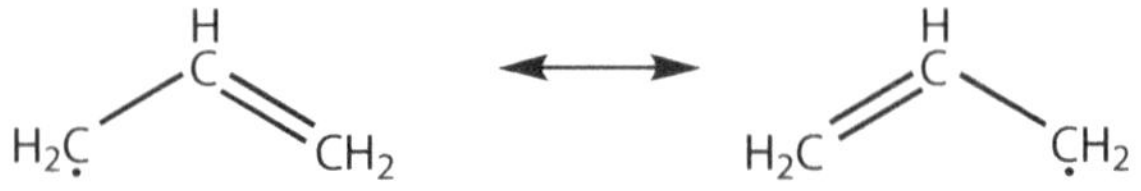

Figure 15.7 Resonance stabilization of free radical.

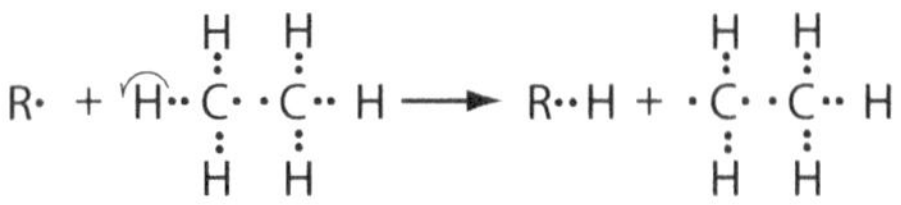

Figure 15.8 Free radical abstraction.

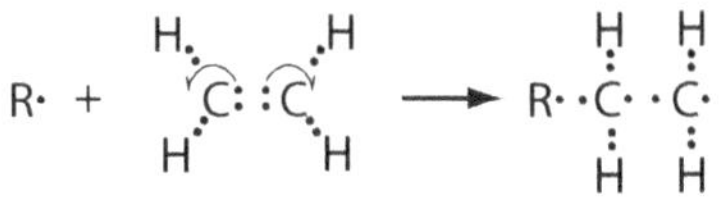

Figure 15.9 Free radical addition.

Figure 15.10 Superoxide ion.

from a double bond. In this position the electrons can shift, giving a resonance form (see Section 9.2, Box on "Energy from ATP") with the unpaired electron on a different carbon atom (Fig. 15.7). Resonance allows the instability of the unpaired electron to be divided among more than one carbon atom. As a result of resonance, an unpaired electron can be produced on one atom, but the subsequent reaction can center about another atom. This can give rise to some chaotic chemistry.

The typical reaction of a free radical is to attach to an atom in another molecule and to grab an electron from its bond. The bonding pair is redistributed with one electron pairing up with the original free radical and the other forming a new free radical. If the bonding pair was a single bond, the bond breaks and an atom is transferred, a process called free radical abstraction (Fig. 15.8). If the bonding pair was part of a double bond, only one bond breaks and the radical joins the group in a process called free radical addition (Fig. 15.9).

The first step in the reduction of oxygen is the formation of the superoxide ion (Fig. 15.10), O_2^-: $O_2 + e^- \rightarrow O_2^-$. The superoxide ion has an odd number of electrons giving it an unpaired electron that makes it very reactive. In beer, the superoxide ion acts as a base, accepting a hydrogen ion from the hydronium ion, H_3O^+ : $O_2^- + H^+ \rightarrow HO_2^\cdot$. The product $HO_2^\cdot$ is the hydroperoxyl (also called perhydroxyl) radical; it is even more reactive than the superoxide ion. The hydroperoxyl radical can initiate reactions with lipids and other compounds in beer, or it can react to give other reactive oxygen species.

The scheme in the box on "Oxidation of Linoleic Acid" shows a simplified version of a possible sequence starting with linoleic acid, the most abundant lipid in barley. The sequence is initiated by the hydroperoxyl radical abstracting a hydrogen atom from carbon-11 between the two double bonds, producing a new free radical and hydrogen peroxide. There are two additional resonance forms of the free radical, one with the unpaired electron on carbon-9 and one with it on carbon-13. The sequence continues with the addition of an oxygen molecule at carbon-9. The resulting peroxide radical abstracts a hydrogen atom from hydrogen peroxide, regenerating the hydroperoxyl radical and forming a lipid hydroperoxide, which has no unpaired electrons. The hydroperoxide is unstable; it rearranges and breaks apart, giving a carboxylic acid with a carbonyl group at the end, and trans-2-nonenal, the nine-carbon aldehyde associated with staling in beer. Once the reaction is initiated, no new reactive oxygen species are needed to keep it going. As long as there is a supply of lipid and of oxygen, the reaction can continue. The carbonyl compounds produced by sequences like this one can contribute to a stale flavor in beer. To prolong the life of beer, brewers try to exclude lipids and oxygen from the brewing process.

The Role of Metal Ions

The formation and interconversion of reactive oxygen species is greatly accelerated by certain metals that are often present at low concentration in beer or wort. Metals that can form two stable ions, like iron (Fe^{2+} and Fe^{3+}) and copper (Cu^+ and Cu^{2+}), are active for interconversions.

- The reduction of oxygen to the superoxide ion:

$$Fe^{2+} + O_2 \rightarrow Fe^{3+} + O_2^-$$

As we discussed above, in an acid solution like beer the superoxide ion exists in the acid form as hydroperoxyl radical $HO_2\cdot$. Two hydroperoxyl radicals can react to give hydrogen peroxide and oxygen. This reaction is typically catalyzed by a metal ion contained in an enzyme called superoxide dismutase, unfortunately known as SOD.

$$2HO_2^\cdot \rightarrow H_2O_2 + O_2$$

- Hydrogen peroxide can oxidize Fe^{2+} forming the extremely reactive and short-lived hydroxyl radical, along with hydroxide ion.

$$H_2O_2 + Fe^{2+} \rightarrow OH^\cdot + OH^- + Fe^{3+}$$

- The Fe^{2+} can be regenerated by transfer of electrons from compounds in the beer to Fe^{3+}. One hypothetical example would be

$$2Fe^{3+} + CH_3CH_2OH \rightarrow CH_3CHO + 2Fe^{2+} + 2H^+$$

- Notice that the oxidation number of carbon in ethanol on the left is -2 and that in acetaldehyde on the right is -1. The iron is being reduced (+3 to +2) and the carbon is being oxidized (-2 to -1).

OXIDATION OF LINOLEIC ACID

Linoleic acid is a fairly abundant fatty acid in grain. It makes up about 1% of the dry weight of barley. The scheme below shows how linoleic acid could undergo reactions with oxygen and reactive oxygen species to give trans-2-nonenal, which imparts a cardboard flavor to stale beer. Trans-2-nonenal can form even if oxygen is rigorously excluded from the package, so its actual origin is more complicated than this scheme.

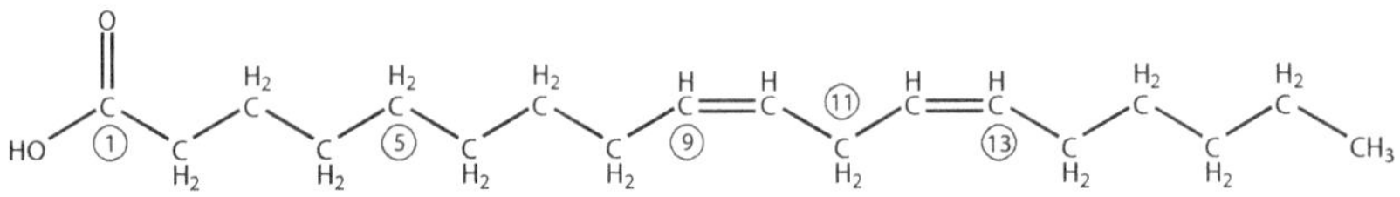

Figure 15.11 Linoleic acid. The numbers in circles identify the carbon atoms in the linoleic acid molecule, starting with the carboxyl carbon.

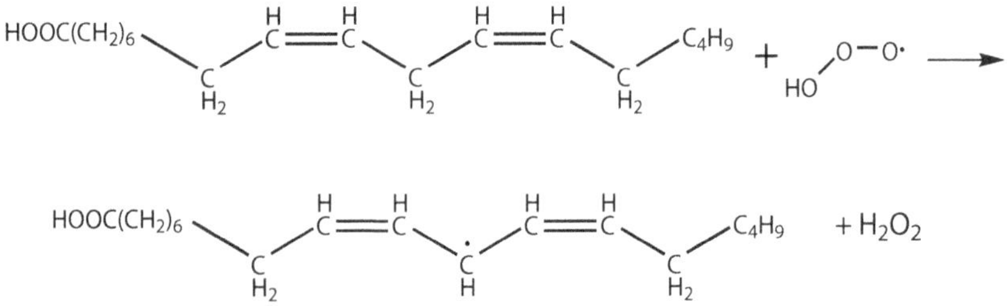

Figure 15.12 Abstraction of hydrogen by the hydroperoxyl radical (HOO·). The product free radical has an unpaired electron on C-11.

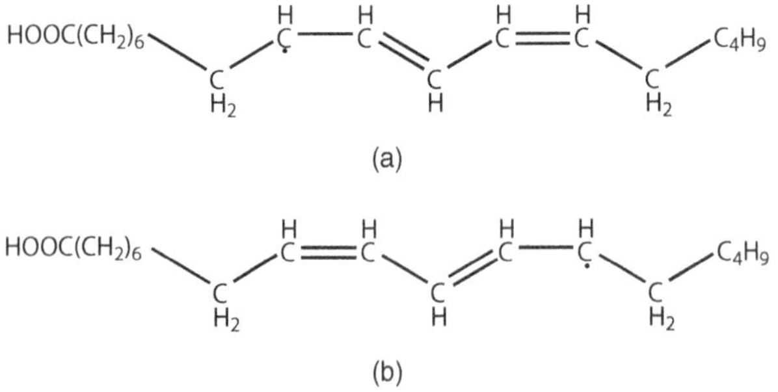

Figure 15.13 Two resonance forms of the free radical. Structure (a) has the unpaired electron on C-9. Structure (b) has the unpaired electron on C-13.

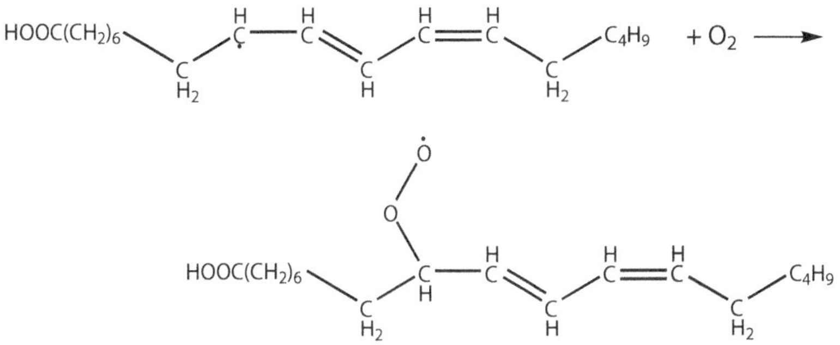

Figure 15.14 Addition of oxygen to the free radical. Oxygen adds to the free radical structure shown in Figure 15.13a to give a peroxyl radical (RCOO·).

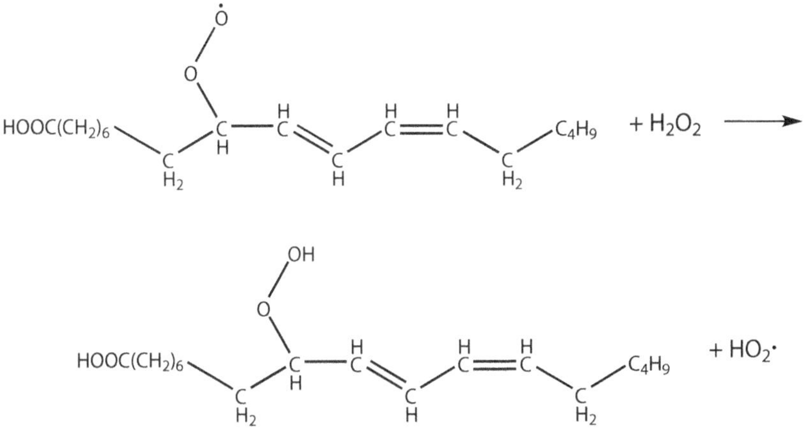

Figure 15.15 Hydrogen abstraction by peroxyl radical. The peroxyl radical abstracts a hydrogen atom from the H_2O_2 generated in a previous step giving a hydroperoxide molecule (ROOH). Hydroperoxyl radical (HOO·) is regenerated so the sequence does not need a fresh source of reactive oxygen species.

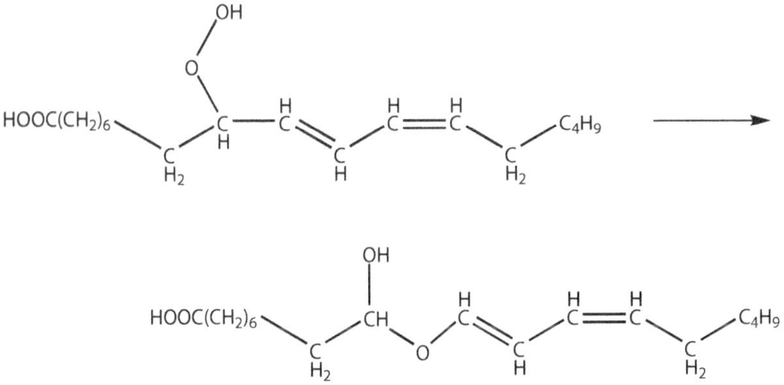

Figure 15.16 Hydroperoxide rearrangement. The hydroperoxide is not very stable; it can rearrange to give a more stable double-bonded hemiacetal.

(continued)

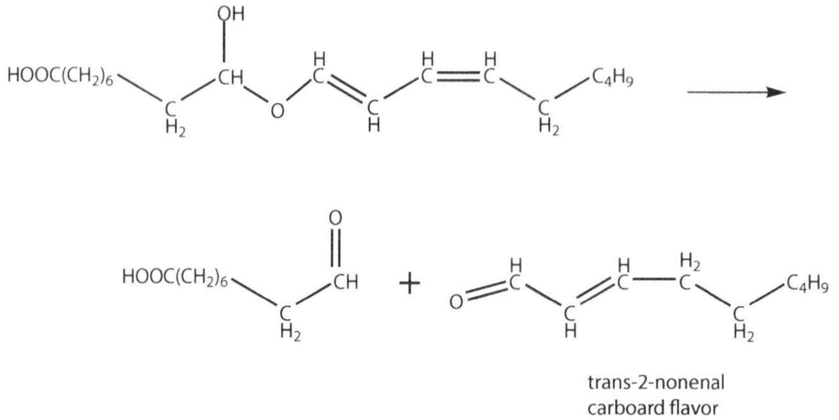

Figure 15.17 Trans-2-nonenal formation. The hemiacetal breaks apart giving trans-2-nonenal as one of the products.

15.3 STALING PREVENTION

If beer could be kept cold from the brewery to the customer, staling would be much less of an issue. The reactions leading to stale flavors in beer, like nearly all chemical reactions, become slower as the temperature is lowered. In general, over a narrow temperature range, the speed of a reaction changes by a certain fraction for each increment of temperature change. Suppose that for a certain reaction, lowering the temperature from 75 to 65 °F results in the reaction going at half its original speed. Lowering the temperature to 55 °F will slow the reaction by half again, that is, to one-quarter of its original speed. At 35 °F the reaction would go only one-sixteenth as fast as it would at 75 °F. If all the staling reactions behaved like this one, beer that would last for 6 months at 75 °F would last for 8 years at 35 °F. Regrettably, the distribution network for beer is not set up to keep the product cold.

Staling resulting from the reaction of oxygen with lipids can be prevented by exclusion of oxygen or lipids. Much of the lipid in beer comes from **trub** and spent grain in the mashing process. Lipid levels can be lowered by effective wort separation. Oxygen is more difficult to control. Air contains 20% oxygen by volume. It is not usually practical to package beer in an oxygen-free environment. Instead, every effort is made to keep oxygen from mixing with the beer during and after packaging. To exclude oxygen during packaging, the equipment is designed to fill the container without splashing. Often the container is filled with carbon dioxide before filling to exclude air. Finally, the beer is often made to foam just before the container is closed to drive off air. Once the container is closed, oxygen exclusion depends on the ability of the packaging to prevent oxygen from crossing the seal. The plastic liners inside the crown caps used to seal glass bottles allow a small amount of oxygen to dissolve in the plastic and migrate into the bottle, a process called **permeation**. The seals on aluminum cans are all metal and permeation is near zero.

Another approach to staling prevention is to reduce the concentrations of free radical reactive oxygen species. Substances that react with free radicals to make less reactive free radicals are called free radical traps, or quenchers. Some components of beer wort, including polyphenols, can act as free radical traps. To assure that these components are not overwhelmed before they reach the package, brewers try to prevent oxygen from mixing with the beer during mashing, boiling, and transfer operations. One very effective free radical trap is sulfite ion, SO_3^{2-}. At the pH of beer, most sulfite ion exists in the acid form HSO_3^-, the bisulfite ion. Many of the reactive oxygen species that can participate in beer staling reactions are deactivated by reacting with the bisulfate ion to make less active free radicals. Ultimately, these react to form the sulfate ion, SO_4^{2-}, in very low concentration, which is essentially harmless. Sulfite ion is formed by yeast during fermentation. Compounds providing sulfite ion can also be used as additives, but customers are resistant to additives in their beer. Sulfite is widely used as an additive in wine, often in the form of metabisulfite ion ($S_2O_5^{2-}$, Campden tablets).

A third way to inhibit free radical reactions is to provide species that bind to metal ions and prevent them from accelerating the formation and interconversion of reactive oxygen species. Various species can bind to certain metals forming covalent bonds in which the species provide both of the shared electrons. The species providing the electrons are called **ligands**. A simple example that is not directly relevant to beer stability would be the reaction of ammonia with Cu^{2+} ions to make a deep blue complex ion (Fig. 15.18). The unshared pair of electrons in ammonia provides both of the electrons in the bond with copper(II) ion. Copper ions tied up in this way are less able to participate in electron transfer reactions.

A compound deriving from grain hulls that may serve as a ligand for copper and iron ions, helping to prevent oxidation, is phytic acid (Fig. 15.19). The oxygen atoms in the phosphate groups can provide several pairs of electrons to an iron or other metal ion forming a strong complex. This keeps the metal ion from participating in the interconversion of reactive oxygen species.

Brewers have done all they can to reduce the oxygen content of packaged beer. The public is unlikely to accept additions of antioxidant materials.

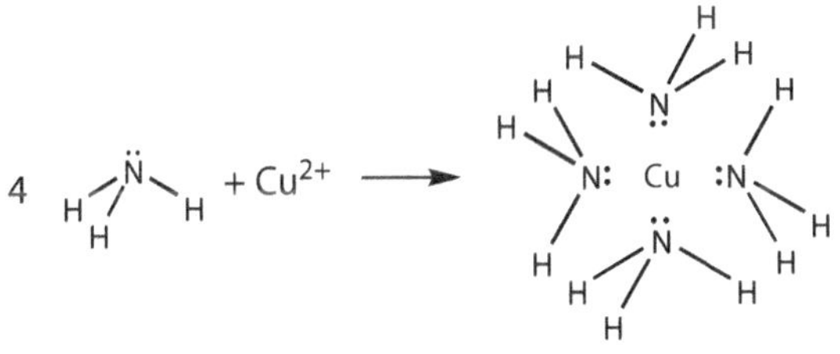

Figure 15.18 Complex formation.

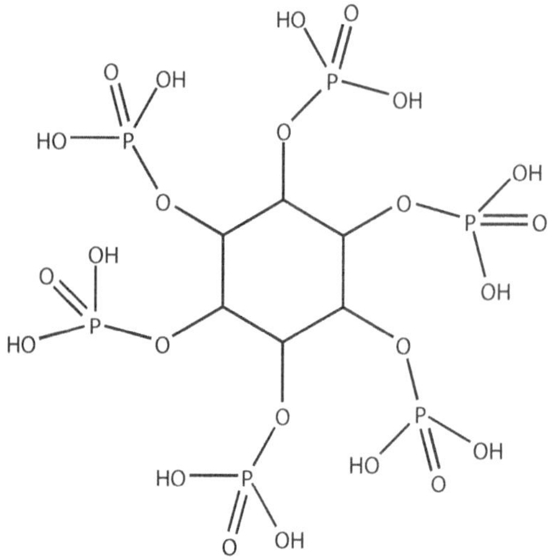

Figure 15.19 Phytic acid.

Improvements in flavor stability may come from brewing processes that take advantage of antioxidants that are naturally available in wort and beer. One way to help these antioxidants survive the brewing process is to exclude oxygen from the mashing and boiling steps. This is likely to be the trend in brewing in the next decades.

BIBLIOGRAPHY

Bamforth, Charles W.; and Parsons, Roy. New Procedures to Improve the Flavor Stability of Beer. *J. Am. Soc. Brew. Chem.*, **1985**, *43*(4): 197–202.

Dalgliesh, C. E. Flavour Stability. *Proc. Congr. Eur. Brew. Conv.*, **1977**, *16*: 623–659, Amsterdam. A review with trenchant commentary. Includes a discussion and graph of flavor changes.

De Shutter, D. P.; Saison, D.; Delvaux, F.; Derdelinckx, G.; and Delvaux, F. R. The Chemistry of Aging Beer. In *Beer in Health and Disease Prevention*, V. Preedy, Ed. Academic Press, 2008, pp. 375–388.

Drost, B. W.; van den Berg, R.; Freijee, F. J. M.; van den Velde, E. G.; and Hollemans, M. Flavor Stability. *J. Am. Soc. Brew. Chem.*, **1990**, *48*: 124–131. Tracks trans-2-nonenal.

Graf, E.; Empson, K. J.; and Eaton, J. W. Phytic Acid. *J. Biol. Chem.*, **1987**, *262*(24): 11647–11650. Phytic acid as an antioxidant.

Kaneda, H.; Kana, Y.; Koshino, S.; and Ohya-Nishiguchi, H. Behavior and Role of Iron Ions in Beer Deterioration. *J. Agric, Food Chem.*, **1992**, *40*: 2102–2107. Iron and reactive oxygen species by electron spin resonance.

Narziss, L. Technological Factors of Flavour Stability. *J. Inst. Brew.*, **1986**, *92*: 346–353. A review with 52 references.

Vanderhagen, B.; Neven, H.; Verachtert, H.; and Derdelinckx, G. The Chemistry of Beer Aging. *Food Chem.*, **2006**, *95*, 357–381. A comprehensive review with 194 references.

Vanderhagen, B.; Delvaux, F.; Daenen, L.; Verachtert, H.; and Delvaux, F. Aging Characteristics of Different Beer Types. *Food Chem.*, **2007**, *103*: 404–412. Sensory and analytical data after one year of storage.

QUESTIONS

15.1. List some flavor changes associated with beer staling.

15.2. Identify compounds associated with catty and with cardboard flavors.

*15.3. For the reaction below, determine the oxidation numbers of each atom in the reactants and products. Indicate what, if anything, is being oxidized and what is being reduced.

$$H_3C\text{—}CH_3 + H_3C\text{—}CH=O \longrightarrow H_3C\text{—}CH(H_2)\text{—}OH + H_2C=CH_2$$

15.4. Write the formulas for the following reactive oxygen species: oxygen, superoxide ion, hydroperoxyl radical, hydrogen peroxide, and hydroxyl radical.

15.5. Draw the resonance forms that result from free radical abstraction of a hydrogen atom from the underlined carbon atom:

$$H_3C-\underline{C}H_2-CH=CH-CH_2-C_2H_5$$

15.6. What is the role of iron and copper ions in beer aging?

15.7. Explain the role that phytic acid could play in enhancing flavor stability of beer.

15.8. Discuss three strategies for extending the shelf life of packaged beer.

15.9. What is the role of superoxide dismutase?

15.10. What is a free radical? What are the two characteristic reactions of free radicals?

*15.11. Determine the oxidation number of each atom in the following species. Don't forget to include the unshared electrons in the Lewis structures.

(a) $H_3C-C(=O)-C(=O)-OH$ (b) $H_3C-O-C(=O)-CH_3$ (c) $H-O-O\cdot$

The author, Roger Barth, brewing beer.

CHAPTER 16

BREWING AT HOME

Brewing can be a great hobby for people who like to enjoy beer with friends and who take pleasure in making their own food and drink (Fig. 16.1). It is not an efficient way to save money on beer. Homebrewers usually brew ale rather than lager beer because ale takes less time and does not require low temperature fermentation. There are two main home brewing techniques, **extract** brewing and **full mash** brewing. In addition, there is a hybrid technique called **partial mash** brewing. The **extract** brewer buys malt extract, adds it to water, boils, chills, ferments, and bottles. The earlier steps—crushing malt, **mashing**, and wort separation—are done by the manufacturer of the extract. The partial mash technique involves mashing some grain, often in a cloth bag, then boiling the resulting wort with added malt extract to provide more fermentable sugar. The partial mash technique takes less space and time and uses less equipment than the full mash technique. The convenience of malt extract comes at a price. Extract is more expensive than grain malt and it is perishable, so it is not usually practical to buy extract in large quantities for less money. Beer made from malt extract is generally somewhat darker than beer made from the equivalent malt by full mash brewing. Some people prefer the taste of full mash beer, but beers containing extract have done well in competitions. We begin our discussion with full mash brewing because it starts earlier in the brewing process. Most home brewers make their first beer with extract; some never move to full or partial mash.

The procedures given here are by no means unique or authoritative. Every brewer has his or her way of doing things dictated by his or her own physical

---

The Chemistry of Beer: The Science in the Suds, First Edition. Roger Barth.
© 2013 John Wiley & Sons, Inc. Published 2013 by John Wiley & Sons, Inc.

Figure 16.1 Brewing.

layout and circumstances. These are my procedures for a 5 gallon (19 liter) batch of ale. It is important to keep in mind that beer is food. Everything that it goes into or that goes into it must be clean.

16.1 SAFETY ISSUES

Brewing beer involves dealing with large volumes of hot liquid and handling heavy objects. Attention to safety is essential.

- Containers of hot liquids in volumes greater than 3 gallons (11 liters) must not be carried about. Provision must be made to chill them in place or to move the liquid through tubing.
- All pots used to handle hot water or wort must have sturdy handles. If the handles show any sign of loosening or cracking, the pot must be discarded.
- Food buckets full of liquid should not be lifted by the bail (handle). The thick ring of plastic that holds the bail affords a safe grip. A bucket full of beer can weigh more than 50 pounds (23 kg).

- A full bucket or fermenter should not be lifted before checking for a clear path to carry it and a clear, sturdy surface to put it on.
- Hands and clothing must be kept clear of the mill, motor, and drive belts.
- Do not bottle beer until you are sure that it is fermented out. Carbon dioxide produced by fermentation can give a dangerous buildup of pressure in bottles, leading to shattering, flying glass, and injuries.

16.2 FULL MASH BREWING

Equipment

Mill Many suppliers will mill malt for customers. This has the disadvantage that malt will only stay fresh for a few weeks after milling. Grain mills for home brewing (Fig. 16.2) are usually set up to be hand cranked (because of liability issues). Crushing 10 pounds of malt with a hand cranked mill is seldom attempted more than once. Some brewers use an electric drill to turn their mill and some attach a motor. Keep in mind that the function of a mill is to crush

Figure 16.2 Malt mill.

Figure 16.3 Brew pot.

anything that comes between its rollers. Make sure that "anything" does not include your fingers. The mill should be away from other brewing operations because it makes dust that can be bad for the beer. You need a couple of clean, dry 3 gallon buckets to handle the grain.

Scale Beer recipes usually give the weight of each type of malt for a 5 gallon batch. A scale that can weigh up to 5 pounds to the ounce is good to have. A **tare** button that puts the reading back to zero is a convenient way to automatically subtract out the weight of the container.

Big Pot We are talking about a BIG pot with a lid (Fig. 16.3). Seven and one-half gallons would be about right. A pot with a spigot is ideal. An inexpensive choice is an aluminum pot of the sort sold for frying a turkey. There must be no trace of oil or soap in the pot or the beer will have no head.

Stove Boiling beer takes a lot of heat. A full-sized kitchen stove or a high capacity propane burner is needed. Propane is not safe to use indoors.

Mash Tun My mash tun is an insulated water cooler (Fig. 16.4) of the type used to pour Gatorade® on coaches' heads at the conclusion of football games.

Figure 16.4 Mash tun.

This is known as a Gott® cooler or a Rubbermaid® cooler. The 5 gallon size is about right. Some brewers use a rectangular picnic cooler for the same purpose. The push-button valve that comes on the cooler is replaced with a cooler conversion kit available at home brew supply shops. The conversion kit provides a stainless steel ball valve and fittings. Insulated vessels have the disadvantage that the only way to raise the temperature is to add hot water or hot wort.

My mash tun is equipped with a **false bottom**, which is a plate with holes to hold up the grain while the wort flows out (Fig. 16.5). The false bottom has a hose barb that connects to the valve with a short length of heat-resistant tubing. Some brewers use a rolled up tube of stainless steel screen, and some use copper tubing with slots instead of a false bottom.

Another approach is to use a large pot fitted with a valve and a false bottom. This allows the temperature to be raised by applying heat. It is possible to use the brewing kettle for this purpose, but then you need another vessel to hold the hot wort while you wash out the kettle after mashing.

Thermometer The temperatures of the **strike water**, the mash, and the **sparge** water need to be measured accurately. A dial or digital thermometer is best. Using a liquid-in-glass thermometer is just asking for trouble. The accuracy can be checked in water with lots of ice and constant stirring (32 °F = 0 °C) and boiling water (212 °F = 100 °C). Some brewing pots are available with a dial thermometer built in. When reading the thermometer, don't let it touch the sides or bottom of the container.

Figure 16.5 Cooler with false bottom.

Heat-Resistant Tubing The wort will be at about 150 °F when it flows from the mash tun to the kettle. Heat-resistant food-grade tubing is opaque white, unlike regular food-grade tubing, which is clear.

Big Spoon A 21 inch (53 cm) stainless steel spoon is needed to stir the mash and kettle. The spoon needs to be long enough to reach the bottom of the kettle with enough sticking out so you can hold it without burning your knuckles.

Wort Chiller A wort chiller can be made from about 30 feet (9 m) of $\frac{3}{8}$ inch (9.5 mm) diameter copper tubing bent around a cylinder to give loops that fit into the kettle. The two ends project well clear of the pot and slope down away from the pot so any leaks will not drip into the wort. Hoses are led from and back to the sink. The chiller should be used while the pot is still sitting on the stove, to avoid moving a pot of boiling hot wort. The water returning to the sink is hot, so the return hose (Fig. 16.6, foreground) is heat resistant. The hoses connect to the copper tubing with hose clamps. The cold water line goes on a fitting in the sink, also with a hose clamp. A hose clamp is a loop of metal that can be tightened with an adjusting screw. The size has to fit the outside diameter of the hose it will go on.

Jar An 8 ounce (250 mL) glass jar covered with a bit of aluminum foil is convenient to hydrate dry yeast.

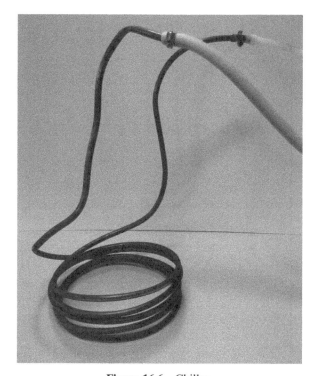

Figure 16.6 Chiller.

Hydrometer The usual way to measure carbohydrate content of wort is with a hydrometer (Section 10.1). Most beer/wine hydrometers have three scales: specific gravity, Balling, and potential alcohol. In this book we refer to the specific gravity scale. Most hydrometers are designed to be used with a liquid at 15 °C (59 °F). For temperature corrections, see Table 10.1.

Sauce Pan A 1½ quart sauce pan is useful.

Rubber Gloves Gloves protect the hands from sanitizing solutions.

Fermenting Bucket The standard **fermenter** is a plastic food bucket whose lid has a hole fitted with a rubber washer, called a grommet, for the fermentation lock. Food buckets have a tough ring below the rim that serves as a grip (the handle, or bail, may not be strong enough to support a full bucket). A 6½ gallon (24.6 liter) bucket is minimal, the 7.9 gallon (30 liter) size is better to avoid foam going up into the fermentation lock. A bucket with a 1 inch (25.4 millimeter) hole drilled in the side centered about 1½ inch (4 cm) from the bottom can accept a spigot, which is very convenient. There are some who insist that fermentation must be done in a glass **carboy**, which is a huge glass

Figure 16.7 Carboy with Brew Hauler.

bottle. Glass, being nonporous, does not retain any odor from previous batches. Carboys have their disadvantages. The carboy has a single narrow opening through which it has to be cleaned, rinsed, filled, and emptied. There is no good place to grip the carboy except the neck. A device made of polymer webbing straps called a Brew Hauler (Brew Hauler, Inc.) is helpful (Fig. 16.7). A 5 gallon glass carboy weighs 11½ pounds empty, compared with 3¾ pounds for a 7.9 gallon bucket with lid and valve. Add that to the weight of 5 gallons of wort, about 44 pounds, and the filled carboy weighs over 55 pounds (25 kg), which not everyone can safely lift by the neck of a bottle. A batch of beer will fill the carboy nearly to the top, leaving no room for foaming. In addition, beer is sensitive to light, so the carboy has to be kept in the dark or shrouded. By contrast, a 7.9 gallon bucket weighs less than 48 pounds (22 kg) full, has a good grip, has space for foaming, and does not admit light. If you bump into a doorknob carrying your full glass fermenter to the garage it will probably shatter and you will have to flee to the Aleutian Islands and change your name to avoid the retribution of your roommates, spouse, or landlord. Breakage is much less likely with plastic. Nonetheless, if the beer is going to be in the fermenter for a long time, the risk of picking up flavors from the plastic may make it wise to ferment the beer in glass. You will need a one-hole rubber stopper that fits the carboy to accommodate the fermentation lock.

Spigots/Siphons Some brewers use **siphons** instead of spigots. Siphons are messy and inconvenient. The usual type of **spigot** will attach to a 1 inch hole in a food bucket. It takes one or two big washers on the outside and a plastic nut on the inside (in the beer). The outlet can accept $\frac{3}{8}$ inch plastic tubing. A type with a lever whose position shows whether it is closed or open can avoid some messy accidents. You need two spigots, one for the fermenter and one for bottling.

If your vessel does not have a hole to accommodate a spigot, you will need a siphon. Any piece of sanitized tubing can serve as a siphon. In order for a siphon to work, the outlet end must, at all times, be lower than the surface of the liquid in the vessel being emptied. The trick is to fill the tube with beer or water, cover one end, put the other end in the liquid in the vessel to be emptied, put the covered end into the receiving vessel, then uncover the tube. There are self-starting siphons that have a hand-operated pump to fill the tubing after it is in place. It is possible for the brewer to start the siphon by sucking the liquid into the tube by mouth (gargling with vodka first in an effort to sanitize one's mouth has been mentioned). This is a poor idea. Take care not to let the inlet end of the siphon drop into any solids at the bottom or they will end up in the receiver.

Fermentation Lock This is a little plastic tube with an S-curve that gets filled with water and lets gas out but no air into the fermenter. It fits in the grommet in the lid of the fermenter, as shown in Figure 16.8.

Bottle Washer This device attaches to the sink and gives a strong jet of water when you push a bottle down on it. Some can wash two bottles at once (Fig. 16.9). The plastic kind works well, but it can't be used with hot water. The bottle washer saves water and time, and gets the bottles clean.

Bottling Bucket A bottling bucket is another food bucket that can accept a spigot. The 6½ gallon size is good. A lid makes it easier to sanitize without getting solution all over.

Bottle Filler A bottle filler is a plastic tube with a valve at the bottom (Fig. 16.10). It opens when the tube touches the bottom of the bottle and closes when you take the bottle down.

Tubing You need enough food grade $\frac{3}{8}$ inch plastic tubing to go from the fermenter spigot (or siphon) to the bottom of the bottling bucket.

Bottle Capper This is the device that seals crown caps to bottles. See Figure 16.16.

Notebook Keep records so you know what worked well and what didn't. Record the date, batch number, yield, recipe, and how it turned out.

Figure 16.8 Fermentation lock.

Full Mash Supplies This is a generic list of supplies for a 5 gallon (19 liter) batch of all-malt beer flavored only by hops. Prices come from Web sites of continental U.S. suppliers and can vary widely.

Water Untreated tap water may not be best for brewing. Passing the water through a carbon filter that fits on the tap removes the disinfectants, which can be a big improvement. If your water is very hard or does not taste good, you can treat it or buy spring water or purified water. Spring water may be available at a low price if you bring your own bottles. For full mash brewing you will use about 8 gallons.

Malt Most recipes call for 9 to 11 pounds (4 to 5 kilograms) of malt. At least half of this should be base (light-colored) malt. Malt prices range from $0.75 to $2.00 a pound.

Sanitizer Any equipment that will be in contact with the wort or beer after the boil must be **sanitized**. **Sanitizers** are chemicals applied to objects to kill microbes. Some sanitizers, like bleach, must be thoroughly rinsed off so they

Figure 16.9 Bottle washer.

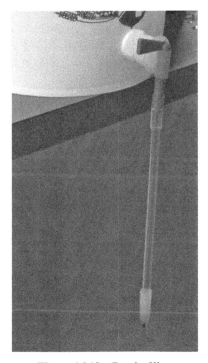

Figure 16.10 Bottle filler.

TABLE 16.1 Sanitizers

Name	Use per Gallon	Use per Gallon	Cost per gallon (U.S. currency)
Bleach	0.5 fl oz	15 mL	1¢
BTF® iodophor	0.1 fl oz	3 mL	5¢
Easy Clean®	1 Tbs	18 g	13¢
One Step®	1 Tbs	18 g	19¢
IO Star®	0.2 fl oz	6 mL	7¢
Star San®	0.2 fl oz	6 mL	10¢

do not get into the food. Others, called "no-rinse" or "leave in" sanitizers can, if used according to instructions, be left on surfaces used for food or beer without affecting the safety or flavor. No-rinse sanitizers are very convenient for bottling because of the large number of items that need to be sanitized. Avoiding a final rinse saves time and water. Unscented laundry bleach is effective and inexpensive, but it must be rinsed out. The active ingredient in bleach is the hypochlorite ion, OCl^-. Other sanitizers commonly used in brewing are Easy Clean® and One Step® (sodium percarbonate: releases hydrogen peroxide, H_2O_2), iodophor and IO Star® (iodine), and Star San® (phosphoric acid). These do not need to be rinsed, which makes them easier to use than bleach. If your rinse water may have microbes, use a no-rinse sanitizer. There is a comparison of sanitizers in Table 16.1. It is possible to sanitize glass and metal with heat instead of chemicals. One hour at 350 °F (177 °C) is about right. Glass should be heated and cooled slowly to avoid introducing strains that can lead to failure. Don't try to sanitize plastic with heat.

Hops Hops come as pellets in sealed bags of 1 ounce (28 grams) pre-weighed. There are a number of varieties that impart different flavors and aromas to the beer. Hops should be kept in the freezer. Most recipes call for 1 to 3 ounces of hops. A 1 ounce bag of hops costs about $2.00.

Yeast Brewing yeast is available in dry and "liquid" (actually a **slurry**) form. Only authentic brewing yeast from a brewing supply store is suitable. Baking yeast and "brewer's yeast" from a health food store are not satisfactory. There are dozens of strains of yeast, each of which imparts a particular character to the beer. At one time, liquid yeast was clearly superior to dry yeast in terms of the level of bacterial contamination and in terms of the variety of strains available. Dry yeast is now of very good quality and the variety, although not equaling that of liquid, is sufficient to make many ale styles. The variety of dry lager yeast is still very limited. The advantages of dry yeast are that it keeps for a long time in the refrigerator, it can be shipped without refrigeration, some brands are inexpensive, and it has enough cells in a package to start the fermentation quickly. Liquid yeast works best when a starter is prepared a couple of days ahead of time. A pack of yeast ranges from $1.25 for some brands of dry yeast to $7.00 or more for liquid yeast.

16.3 FULL MASH BREWING PROCEDURE

Strike Water

The **strike water** is the first water added to the malt. Some recipes call for starting at a lower temperature and then increasing (two-step mash) and some call for taking the temperature all the way up at once (one-step mash). Starting at a lower temperature extracts more sugar from the grain and gives more complete fermentation. Some say a two-step mash gives clearer beer. A one-step mash leaves more unfermented carbohydrate in the beer, which is said to provide body. The strike water temperature is not the same as the **mash-in** temperature, that is, the temperature after the water and malt are mixed. When the hot water is added to the malt, the water uses some of its energy to warm up the malt and the mash tun. The mash-in temperature is generally 10 to 15 °F (5.5 to 8.3 °C) lower than the strike water temperature. The first step in full mash brewing is to rinse the kettle well, add 2 gallons of water for a two-step mash or 3 gallons for a one-step mash. Heat the water to the strike temperature, which would be around 155 °F (68 °C) for a two-step mash and 163 °F (73 °C) for a one-step mash, depending on the recipe.

Grist

While the water is heating, weigh out the malt into a bucket according to the recipe. If you are using whole malt, crush it into another bucket.

Mash

Rinse the mash tun, its lid, and the false bottom well. Attach the fitting on the false bottom or screen to the valve on the cooler. When the strike water is ready, add enough to the empty mash tun to cover the false bottom so there is no air bubble under the grain. Then pour in the grain carefully to minimize dust. Add the remaining strike water gently to minimize pick-up of air. Stir the mash well and measure its temperature while stirring. If the temperature is a lot too high, it can be brought down with cold water. If the temperature is not extreme, it is best to reserve judgment, close the lid for a few minutes, then stir and measure again. It is difficult to get a reliable temperature measurement in a thick mash. If the temperature is too low, hot water can be added, or some mash can be scooped out with a sauce pan, boiled, and put back in. If the mash is a lot too hot for a long time, the enzymes that convert starch to sugar will be destroyed, so there won't be much sugar for the yeast to ferment. For a two-step mash, add a gallon of boiling water after 15 minutes to bring

Figure 16.11 Vorlauf. (See color insert.)

the temperature to the second step. Cover the mash tun and put it on a countertop to sit for about 40 minutes.

Sparge Water

While the mash is mashing, heat 4 gallons of water to 172 °F (78 °C). It is convenient to have a second pot to do this so the kettle is free.

Vorlauf

After about 40 minutes, open the mash tun valve part way to let some of the wort into the sauce pan. It will be very cloudy. Pour the wort back into the mash tun (Fig. 16.11). Repeat this until the wort runs clear. This process is called **vorlauf** (German: forerun). Don't open the valve too far. If the wort runs too fast the grain bed can become compacted and the flow may be blocked.

Sparge

Run heat-resistant tubing from the valve on the mash tun to the bottom of the kettle (or to a food bucket if the kettle is used for sparge water). Let the wort down slowly. Avoid bubbling and splashing because oxygen is not good for the beer at this stage. When the level of the liquid in the mash tun reaches the top of the bed of grain, add a saucepan of sparge water (Fig. 16.12) at about 172 °F (78 °C). Keep running off the wort and adding sparge water until all the sparge

Figure 16.12 Sparging. (See color insert.)

water has been added to the mash tun, then let the rest of the wort down. If the wort is in a food bucket, let it down into the kettle through tubing without splashing.

Boil

Stir the wort and take a sample of around $\frac{1}{2}$ pint (250 milliliters) into a saucepan. Cover the kettle part way and begin heating. When the sample in the sauce pan cools to room temperature, pour it into the hydrometer jar. Put the hydrometer in gently and take a reading where the liquid meets the stem of the hydrometer (see Fig. 10.2). It is likely that the reading will be higher than the target original gravity for the recipe. Add water to the pot until the specific gravity is correct or until the wort comes to about 2 inches (5 cm) from the top of the pot, whichever is less. Keep the kettle mostly covered but with an opening. During heating, stir the wort occasionally. When froth forms on top of the wort, skim it off to avoid boiling over. While the wort is heating up, prepare sterile water according to the next step: *Hydrate Yeast*. When boiling begins, temporarily turn off the heat (to avoid boiling over). Carefully, with stirring, add **boiling hops** (also called **bittering hops**) as called for in the recipe. Resume heating. A good rolling boil should be maintained. Most recipes call for additional hops, sometimes called **aroma hops**, to be added at various intervals during the boil. When the boiling time (usually 1 hour) is over, take out a sample, cool it, and check the specific gravity again. If it is too high, add water.

Hydrate Yeast

Prepare a jar of sterile water by rinsing the inside of the jar and an aluminum foil cover with boiling water. Then add about 3 ounces (100 mL) of boiling water, cover it with the foil, and let it sit until cool. When the water is completely cool (below 90 °F = 32 °C), sprinkle the yeast onto it and cover it with the foil. After 10 minutes, stir the yeast to get it all wet. The yeast should soak for 15–30 minutes. Anything that goes into or touches the yeast, including thermometers and stirring spoons, must be sanitized to prevent contamination.

Sanitize Fermenter

There are three levels of cleanliness: clean (free of adhering dirt), sanitized (treated to lower the level of microbes), and sterile (free of any live microbes or viable spores). Sanitizing is a treatment to bring the level of microbes (bacteria and fungus) to tolerable levels. It is not the same as sterilization. For sanitizing to be effective, the surfaces must already be clean, that is, free of dirt or grease in which the microbes can hide from the sanitizing treatment. While the wort is boiling, sanitize the fermenter spigot by soaking it in sanitizer solution. The sanitizer solution must go into and through the interior of the valve. Wear gloves when handling sanitizer; after a while it irritates the skin. Rinse the fermenter bucket to make sure it has no dust or soap (it should already be clean). Install the spigot on the fermenter. The washers go on the shank (threaded part); the shank goes into the hole from the outside; and the nut is screwed on tightly from the inside. If the outlet of the valve sticks down to table level, turn the valve to the side, always moving it in a clockwise direction so it doesn't come loose. Close the valve and add about a gallon of water and the correct amount of sanitizer concentrate (for bleach, about ½ ounce = 15 mL). Put a clean fermentation lock into the sanitizer and make sure it gets solution on the inside. Cover the fermenter and its grommet hole, and shake and roll the fermenter to get sanitizer on all inside parts. Toward the end of the wort boil, let the solution out through the valve, recover the fermentation lock, and put it in a sanitized place like the bowl used to sanitize the valve. Rinse the fermenter well to clear it, the lid, and the valve of all bleach (if used). Put the lid on the fermenter and set it aside in a clean place.

Chill

Once the wort cools, it becomes vulnerable to contamination by wild yeast and bacteria. This is the main cause of problems in brewing. Anything that touches the wort or anything that the wort is going to touch must be sanitized.

Figure 16.13 Aerating.

Rinse the chiller well, connect it to a cold water faucet, and put it into the wort. The hot wort will sanitize the chiller at once. Run cold water through the chiller at a moderate rate. Stir the hot wort regularly by moving the chiller up and down. Continue chilling until the pot is cool to the touch.

Aerate

Put the sanitized fermenter, with its valve closed, on the floor. Run the chilled wort from the kettle into the fermenter with lots of splashing to get air into the wort (Fig. 16.13). The boiling process removes all dissolved oxygen from the wort. The yeast needs oxygen to get started. For more effective aeration, filtered air or pure oxygen can be pumped into the wort using a diffuser with tiny holes to make small bubbles.

Pitch

Dry the outside of the jar in which the yeast is hydrating and slosh it to make sure the yeast is not all stuck to the bottom. Pour the yeast slurry into the fermenter (Fig. 16.14). It is not important to get every drop. Putting yeast into wort is called **pitching**. Cover the fermenter tightly. Carry the fermenter to a part of the house where it does not get too hot or cold. Rinse the fermentation

Figure 16.14 Pitching.

lock and leave water in it so that each arm has about 1 inch (2.5 cm) of water. The fermentation lock goes into the grommet on the lid of the fermenter. Bubbling is usually seen in the fermentation lock within 12 hours for dry yeast. Liquid yeast can take as much as two days to start. Leave the fermenter undisturbed for about a week. If, for any reason, you need to move the fermenter, remove the fermentation lock first so you don't get water of dubious microbiological quality into your beer.

Secondary Fermentation

If the beer is going to be in the fermenter for more than two weeks, it may be desirable to transfer it to a secondary fermenter. This allows the fermentation to continue without a mass of dead yeast cells. The transfer should be done when the primary fermentation has slowed down, but before active fermentation is completely finished. Sanitize a 6.5 gallon food bucket, a lid, a fermentation lock, a piece of tubing, and a spigot. Rinse the sanitized bucket of all bleach, rinse the tubing, put the bucket under the fermenter, and lead the tubing to the bottom of the bucket. Let the beer run into the bucket slowly. When the end of the tubing is covered, you can run the beer more quickly. Try to avoid getting any air in the beer. Cover the bucket and install the fermentation lock. Allow the bucket to stand until fermentation is complete. Some brewers prefer to use a glass carboy for secondary fermentation. A carboy uses a one-hole rubber stopper to hold the fermentation lock instead of a lid with a grommet.

16.4 EXTRACT BREWING

Extract Brewing Equipment

For extract brewing you need:

4 gallon pot (or larger)
Stove
Big spoon
Everything listed for full mash from *rubber gloves* on

Extract Brewing Supplies

Water Six gallons of water is needed. Not all of the water is boiled with the wort. If there is any chance of microbes in the water, 3 gallons of it should be boiled and chilled and put into the sanitized fermenter before wort boiling.

Malt Extract Most recipes call for 5–8 pounds of extract. Extract comes as syrup (liquid malt extract, LME) or as powder (dry malt extract, DME). The recipe will indicate what to get. Dry malt extract is more concentrated, so less is needed.

Other Supplies All the supplies used for full mash from *sanitizer* on.

USING LIQUID YEAST

Many brewers believe that liquid yeast makes better beer. There is no question that there are more varieties of liquid yeast available. Liquid yeast is actually a slurry of yeast cells suspended in beer wort. It comes in tubes or plastic bags. The package of yeast must be kept cold until used. A package of liquid yeast has about half as many cells as a package of dry yeast. Also, the cells in liquid yeast are in a dormant condition. For these reasons, it is good to make a **starter**, that is, to grow the yeast in wort to increase the number and vitality of the cells before pitching it into your wort. If you bypass using a starter, you can expect to wait for 12 hours or more for fermentation to start. Whether or not a starter is made, the wort must be thoroughly aerated before the yeast is pitched. Filtered air can be pumped in through a stainless steel air stone. Start

(continued)

the air flow before you put the stone into sanitizing solution, and keep it going until you are finished aerating. It can be impossible to drive water out of the pores in the air stone once it gets in. Some brewers aerate with pure oxygen; care must be taken not to introduce too much.

Making a Starter

Allow the liquid yeast to warm to room temperature. Weigh out 4 ounces (112 g) of pale dry malt extract or 5 ounces by weight of pale liquid malt extract (136 g or 94 milliliters or 3.5 fluid oz), $\frac{3}{8}$ teaspoon (0.6 mL or 0.6 gram) of yeast nutrient, and a magnetic stir bar into a 2 liter stovetop-safe Erlenmeyer flask (conical bottle). Add 3 pints (1400 milliliters) of water and 2 drops of Fermcap S® (Kerry) foam inhibitor. This will give a wort with specific gravity of about 1.030. Cover the flask with a piece of aluminum foil, heat it on a stove, and allow it to boil for 15 minutes. Keep the wort from boiling over. Cool the flask to pitching temperature (about 70°F). Aerate the wort. Put the covered flask on a magnetic stirrer and begin stirring. Sanitize the outside of the container of liquid yeast, shake it up, and pour the yeast into the flask, Replace the foil. Allow the yeast to incubate with continuous stirring for 2–3 days until the final gravity is reached. Turn off the stirrer and let the yeast settle for several hours. Remove as much as possible of the clear liquid from the yeast before pitching. Try to have the starter and the wort at about the same temperature when you pitch to avoid shocking the yeast.

Extract Brewing Procedure

Boil Heat 3 gallons of water to a boil. Turn off the heat and sprinkle dry extract in while stirring to keep it from clumping. A wire whisk may be helpful. If you are using liquid malt extract, pour it in while stirring. The stuff is very thick and sticky. As much as possible should be scraped from the sides of the can. If the LME comes in a bottle, it is helpful to warm the bottle in hot water so it will flow better. Stir the pot well to get everything dissolved and then bring it back to a boil. Boiling proceeds in the same way as for full mash brewing. Any foam should be skimmed to avoid boiling over. When boiling starts, add the bittering hops. Maintain a rolling boil with the pot partly covered. Add hops at times given in the recipe.

Hydrate Yeast Use the same procedure as for full mash brewing.

Sanitize Fermenter Use the same procedure as for full mash brewing.

Chill When the boiling time is over, the wort is chilled. This can be done by putting a bunch of ice in the sink and putting the pot in it, making sure not to allow any ice or sink water into the wort. This is not very convenient, and it is better to use a chiller as explained under *full mash equipment*. Some brewers add food-grade ice directly to the kettle, if there is enough room.

Aerate Add 3 gallons of good water to the fermenter. Wipe the outside of the kettle if it is wet and pour the wort into the fermenter with splashing.

Pitch Follow the same procedure as for full mash brewing.

GETTING STARTED CHEAP

A standard starter kit for home brewing costs $80 to $160, not including the brew kettle and supplies. We will introduce two alternatives, the cheap and the supercheap options. Convenience is being sacrificed for economy. These setups are suitable for extract brewing. Both include a 5 gallon pot. A capper is not included; plastic soda bottles can be used. Avoid bottles from fruit-flavored soda.

Brewing Start-Up Options

CHEAP OPTION

Item	Amount	Each	Net
20 quart aluminum pot	1	$18.00	$18.00
6.5 gallon buckets, drilled	2	$15.00	$30.00
Lid for bucket, drilled	1	$3.00	$3.00
Spigots	2	$3.75	$7.50
Tubing—3/8 inch inside diameter	4 ft	$0.35	$1.40
Fermentation lock	1	$1.10	$1.10
Total			$61.00

SUPER CHEAP OPTION

Item	Amount	Each	Net
20 quart aluminum pot	1	$18.00	$18.00
Generic food buckets with lids	2	$8.00	$16.00
Tubing—3/8 inch inside diameter	6 ft	$0.35	$2.10
Pinch clamp	2	$0.50	$1.00
Grommet	1	$0.50	$0.50
Fermentation lock	1	$1.10	$1.10
Total			$38.70

(*continued*)

The supercheap option is for the real do-it-yourself brewer. Drill a $\frac{3}{8}$ inch hole in one of the lids about 3 inches from the edge to take the grommet. You will need to move wort and beer about with siphons made from the tubing. The pinch clamps will help control the flow. Use one to set the rate and the other for on–off. You may be able to get the buckets for free or minimal cost from a restaurant or bakery, but clean them well. Practice siphoning with water first.

16.5 BOTTLING

Bottling is the same for full mash as for extract. Don't bottle until there has been no bubbling from the fermentation lock for a day.

Bottling Supplies

Bottles One gallon is 128 ounces, or $10\frac{2}{3}$ twelve-ounce (355 milliliter) bottles. Five gallons of beer will fill 53 bottles. It is good to have some extra. Pry-top brown glass beer bottles are best. Twist-off bottles are a little less reliable at making a seal. Some imported beer bottles will not accept American caps. One or two liter plastic soda bottles with caps will work, although the beer will not last as long in them because oxygen can go through the plastic. Water bottles or other plastic or glass bottles for noncarbonated beverages are no good. Glass bottles are quite expensive new (60¢ each), so you will probably want to clean and reuse bottles.

Caps Crown caps are used to seal glass bottles. These are available from brewing supply stores in packages of 144 (a gross). You can get them in colors, which is convenient for keeping track of different styles. Plain caps cost about 2 cents each, colored caps are slightly more.

Corn Sugar This is a form of glucose (also called dextrose). Other sugars are satisfactory substitutes. Don't use honey. Five ounces by weight (140 grams) or six fluid ounces (180 mL) are needed for bottling. This costs about 40 cents.

Sanitizer See Table 16.1.

Bottling Procedure

Sanitize The first requirement is washed bottles. The bottles must be rinsed well before they dry to remove all traces of soap. A bottle washer is a big help. Wash and thoroughly rinse a 6½ gallon food bucket and install a sanitized spigot. Put 2 gallons of tap water and the correct amount of sanitizer in the bucket. Put in enough tubing to go from the fermenter to a bucket on the floor, making sure that the hose fills and the inside gets sanitized. Put in a bottle filler, making sure it gets filled and sanitized on the inside. If you use a piece of hose to connect the bottle filler to the spigot, that also must go into the bucket to be sanitized. Cover the bucket and shake it around to get sanitizer solution on all inside surfaces. Sanitize all surfaces of a big plastic or metal spoon to stir the beer. After a few minutes, take the bottle filler out and install it on the spigot. Use the filler to put about 2 inches (5 cm) of sanitizer solution into each bottle. As you fill the bottles, cover them with a (gloved) thumb and shake them to get sanitizer solution on all inside surfaces. Take the bottle filler off and fill a bowl with sanitizer solution. Put crown caps in the bowl and make sure the insides get sanitizer solution on them.

A no-rinse sanitizer is very convenient for the bottling process. When used exactly according to the manufacturer's directions it avoids a final rinse step both for the bottles and the caps, saving time and water.

Priming Measure out 5 ounces by weight (140 grams) or 6 ounces by volume (180 mL) of corn sugar into a glass cup that can hold a pint (500 milliliters). Add boiling water to give a pint and swirl until the sugar dissolves. Set the sugar water aside in a clean place.

Transfer Beer Take the fermentation lock off the fermenter (otherwise the water will be sucked in when you let the beer down). Take a sample of the beer, taste it, and measure its gravity. If the gravity is as expected, continue. If the gravity is too high, let the fermentation go for a few more days. It is not safe to bottle beer before it is fully fermented; the bottles may shatter when the sugar finally ferments. Rinse the sanitized bucket of all bleach, rinse the hose, put the bucket under the fermenter, and lead the hose to the bottom of the bucket. It can be helpful to use the big spoon to hold the hose to the bottom. Let the beer run into the bucket slowly. When the end of the hose is covered, you can run the beer more quickly. Try to avoid getting any air in the beer. While the beer is running down, pour the corn sugar water into the bucket (it doesn't matter if it is still hot). Stop the flow of beer before any of the yeast and stuff at the bottom of the fermenter gets in. Stir the beer gently, but thoroughly, to get the sugar evenly dispersed.

Bottling Put the bucket with beer (it is heavy) onto a counter or stout table. Put a basin on the floor below. Rinse the bottle filler and install it on the spigot. Empty a couple of bottles (rinse them well unless your sanitizer is no-rinse) and fill them (Fig. 16.15) until the beer comes right to the top (the space taken by the filler will leave enough head space). When you have filled about 12 bottles, place caps (rinsed unless you are using a no-rinse) on the filled bottles. Continue until all the beer is bottled. If your last bottle will not fill all the way, set it aside and don't cap it. You can drink it, of course. Only full bottles should be capped or excessive pressure can result.

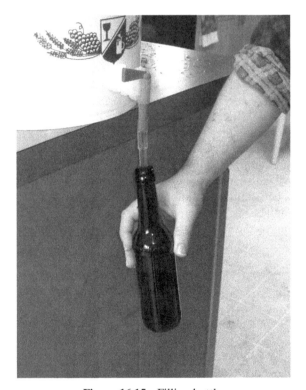

Figure 16.15 Filling bottle.

Capping Put the bottle capper squarely on a covered bottle and squeeze down the levers (Fig. 16.16). Put the capped bottle in an empty case and do the next bottle. Label the case with the date and batch number.

Bottle Fermentation Put the bottles aside in a cool place (65 to 70°F = 18 to 21°C; 45 to 60°F for lager) for one week (2–3 weeks for lager). The sugar added at bottling will ferment to a small amount of alcohol and carbon dioxide. This carbon dioxide is sealed in the bottle, so it carbonates the beer. Yeast will grow and settle out of the beer, so there will be a layer of sediment. The beer will continue to improve for several weeks.

16.6 STARTER BREWING SYSTEMS

There are systems that make 2 gallon (7.6 liter) batches, often with an amber plastic fermenter that has a cute cask-like appearance. These list for about $55 including supplies for one batch. I got one of these as a gift; it was my start in brewing. Refills cost $16 to $25. This is $18 to $28 per case. Compare the cost for a standard 5 gallon batch worked out in the recipes below.

Figure 16.16 Capping.

BREWING LAGER BEER

Brewing lager beer requires that the fermentation temperature be controlled below 60 °F (15.6 °C). This applies to the main fermentation and any bottle fermentation. Lower temperatures give slower reaction rates, so the fermentation and conditioning steps take longer.

The first issue is temperature. Room temperature is too high for lager fermentation. One solution is to run the fermentation and conditioning steps in a refrigerator equipped with a temperature controller. A refrigerator with a freezer has the problem of water condensation. A beverage refrigerator is best because it has no freezer. A cheaper approach is to ferment in a cold place and apply heat to set the fermentation temperature. This will only work if the cold place can be depended upon to stay colder than the desired fermentation temperature for the entire time of fermentation, which can be several weeks. Heat can be provided by wrapping a few turns of heating tape around a glass carboy. Heating tape should not be applied to a plastic fermenter. The heating tape should be

(*continued*)

plugged into a temperature controller with the probe either directly in the beer, or in a tube, called a thermowell, which is open at one end with the closed end extending into the beer.

The second issue is yeast. Lager brewing requires lager yeast. Yeast grows more slowly at lower temperature, so it is best to pitch plenty of yeast in good condition. If you are using dry yeast, it would be good to pitch two packages. If you are using liquid yeast, make a **starter** so the yeast will be pitched at a high level of activity and with adequate numbers of cells.

The third issue is time. Fermentation can take several weeks. Many brewers are reluctant to leave their beer in a plastic bucket for such a long time. It is common to transfer ("rack" in brewing jargon) the beer to a glass carboy after a week or so. After priming and bottling, the beer is held at fermentation temperature for two weeks or so to ferment the priming sugar. After that, the lagering takes place in bottles just above freezing for another few weeks. Typically, bottle conditioned lager beer is ready to drink two to three months after the yeast is pitched for the primary fermentation. If the beer is to be force carbonated in a keg, it is conditioned before carbonation and packaging.

16.7 RECIPES

The full mash recipes are from recent batches that I made. The extract versions are rough translations, but they should work well. These probably will not win any prestigious awards, but most people find them quite drinkable. Costs are from Web-based suppliers in the continental United States, and do not include shipping. They also do not include supplies like sanitizer (less than 10¢ per case) or caps (60¢ a case). It is best to buy from your local homebrew shop. The prices will often be a bit higher, but he or she will pay shipping instead of you. And you can pick her/his brains for free. A case of 24 twelve-ounce bottles comes to 2.25 gallons (8.5 liters).

American Pale Ale Extract

Ingredient	Amount	Cost
Alexander's Pale Syrup can	4 lb	$17.50
Breiss Sparkling Amber DME	2 lb	$10.00
Cascade hop pellets	3 oz	$4.50
Safale US-05 yeast	1 pkg	$3.50
Total cost		$35.50
Cost per case		$15.98

Boil 2.5 to 3 gallons of water. Add the syrup and extract. Add 1 oz of Cascade hops. Boil for 30 minutes then add 1 ounce of Cascade hops. Boil for 20 minutes then add 1 ounce of Cascade hops. After 10 minutes, chill wort to no higher than 85 °F. Add wort to the sanitized fermenter with agitation to aerate the wort and bring the volume up to 5.5 gallons with clean water. Pitch the hydrated yeast. Ferment for 1 week. Final gravity should be around 1.011 (or less).

American Pale Ale Full Mash

Ingredient	Amount	Cost
Pale ale malt	7 lb	$10.50
Crystal or caramel malt	2 lb	$4.00
Cascade hop pellets	2 oz	$3.00
Safale US-05 yeast	1 pkg	$3.50
Total cost		$21.00
Cost per case		$9.45

Mash in with 3 gallons (11 liters) of water at 152 °F (67 °C). Sparge with 4 gallons of water at 172 °F (78 °C). Bring to boil and add 1 oz Cascade hops. After 50 minutes of boiling, add 1 more ounce of Cascade hops. After 10 more minutes add water to adjust the specific gravity to 1.044; chill. Transfer to fermenter with agitation to aerate the wort and pitch hydrated yeast. Ferment for 1 week. The final gravity should be 1.011 or less.

Brown Ale Extract with Specialty Grain

Ingredient	Amount	Cost
Breiss Sparkling Amber DME	5.5 lb	$23.00
Chocolate grain malt, crushed	6 oz	$0.84
Muslin grain or hop bag	1	$0.60
East Kent Goldings pellet hops	2 oz	$3.00
Windsor yeast	1 pkg	$4.50
Total cost		$31.94
Cost per case		$14.37

Put the crushed grain into the muslin bag, leaving some room for expansion, and tie it off. Put 2.5–3 gallons of water into the pot and add the bag of grain. Begin heating. When the water gets to 170 °F (77 °C, a little after it begins to steam) pull out the grain and discard. When the water boils, sprinkle in the extract. Add the hops; boil for 60 minutes. Chill. Put into the sanitized fermenter with lots of agitation to aerate the wort. Add enough clean water to bring volume to 5.5 gallons. Pitch the hydrated yeast. Ferment 10 days. Look for a final gravity of about 1.020.

Brown Ale Full Mash

Ingredient	Amount	Cost
Pale ale malt	6.5 lb	$9.75
Dark Munich malt	1 lb	$3.60
Crystal malt 40 L	0.5 lb	$1.00
Chocolate grain malt, crushed	6 oz	$0.84
East Kent Goldings pellet hops	1 oz	$1.50
Windsor yeast	1 pkg	$4.50
Total cost		$21.19
Cost per case		$9.56

Mash in with 2 gallons of water at 140 °F. Add 1 gallon boiling water and bring the temperature to 158 °F. Sparge with 4 gallons of water at 172 °F (78 °C). Bring to a boil, add hops, and boil for 60 minutes. Add water to adjust to a specific gravity of 1.044. Chill and pitch yeast. Ferment for about 10 days. Final gravity should be about 1.020.

Bock Lager Full Mash

Ingredient	Amount	Cost
German Munich malt	7 lb	$13.00
Carahell malt 10 L	1.5 lb	$3.25
Dark Munich malt	2 lb	$3.70
Pilsner malt	1 lb	$1.80
Hallertau pellet hops	1 oz	$1.80
Tettnang pellet hops	1 oz	$2.00
German Lager Yeast WLP 830	1 pkg	$7.00
Total cost		$32.55
Cost per case		$14.65

Prepare a starter for the yeast 2–3 days before brew day. Strike 2 gallons of water at 160 °F to give a mash-in temperature of about 140 °F. Add 1 gallon of boiling water to bring the mash to 154 °F. Mash for 1 hour. Sparge with 4.75 gallons of water at 172 °F. Boil. Add Hallertau hops. Boil 30 minutes then add Tettnang hops. Boil an additional 30 minutes. Chill to 70 °F. Aerate wort well. Pitch starter. Ferment at 50 °F for 3 weeks or until final gravity is reached. Bottle and keep bottles at 50 °F for 2 weeks, then lower temperature to 32–35 °F for 3 weeks.

BIBLIOGRAPHY

Noonan, Gregory J. *New Brewing Lager Beer*. Brewers Publications, 1996. Outstanding guide for all lager brewers.

Palmer, John J. *How to Brew*. Brewers Publications, 2006. A detailed resource to get you started in homebrewing.

Papazian, Charles. *The Complete Joy of Homebrewing*, 3rd ed. HarperCollins, 2003. The author is the founder of the American Homebrewers Association and president of the Brewers Assocation. His relaxed style is captured in his motto, "Don't worry, have a homebrew!"

QUESTIONS

16.1. Explain the difference between full mash and extract brewing. What are some advantages and disadvantages of each?

16.2. What are the advantages of milling one's own grain?

16.3. Identify the three levels of cleanliness.

16.4. What are some advantages and disadvantages of no-rinse sanitizers?

16.5. What is a siphon?

16.6. What makes brewing lager more complex than brewing ale?

16.7. Compare the advantages and disadvantages of liquid and dry yeast.

GLOSSARY

α-amylase. *See* **alpha-amylase**.

β-amylase. *See* **beta-amylase**.

absorbance. Measure of light absorbed by a sample calculated from 2 − log(%T), where %T is the percentage of light transmitted by a sample.

acetyl coenzyme A. Biological thioester used to transport acetic acid.

acid. Substance whose **molecules** provide hydrogen **ions**.

acidic. Dominated by **acid**.

activated carbon. Highly porous carbon made by treatment of coke or charcoal at high temperature with a material that reacts with it to give a gas. Used to purify water and other **fluids**.

activation energy. Minimum **energy** needed for a chemical **reaction** event.

active site. Location on a **catalyst molecule** or surface responsible for the catalytic **reaction**.

addition. Chemical **reaction** in which the second or third **bond** in a multiple bond breaks and **groups** bond to each end of the broken bond.

adenosine diphosphate. ADP. Adenosine ring with two phosphates, the low **energy** form of ATP.

adenosine triphosphate. ATP. Adenosine ring with three phosphates, the molecule that carries **energy** in cells.

adjunct. Source of fermentable material other than **malted** grain.

The Chemistry of Beer: The Science in the Suds, First Edition. Roger Barth.
© 2013 John Wiley & Sons, Inc. Published 2013 by John Wiley & Sons, Inc.

ADP. *See* **adenosine diphosphate**.

aerobic. Using oxygen.

alcohol. 1. **Organic** compound with the –OH **group** as its principal **functional group**. 2. **Ethanol**.

alcoholic beverage. Liquid containing **ethanol** for human consumption.

alcoholism. Addiction to **ethanol**.

aldehyde. Compound with a C=O **group** bound to one or two hydrogen atoms.

aldose. Monosaccharide originating as an **aldehyde**.

ale. 1. **Beer (1)** made with **gruit**. 2. Beer fermented at a temperature above 60 °F.

aleurone layer. Thin layer of living cells beneath the **seed coat**. Produces enzymes when the plant **embryo** germinates.

alkaline. *See* **basic**.

alkalinity. Measure of the amount of **carbonate** and bicarbonate in water.

alkaloid. Naturally occurring **amine**.

alkane. Organic compound with only carbon and hydrogen and only single **bonds**.

alkene. Hydrocarbon with one or more C=C double **bonds**.

alkyl group. Part of a **molecule** including only C and H **atoms** with only single **bonds**.

alpha acid. Type of compound in **hops** that, after **isomerization**, gives hop bitterness.

alpha-amylase (α-amylase). Enzyme that **hydrolyzes** 1→4 **glycosidic links** on a **starch** chain anywhere along the length of the chain.

amide. Carbonyl compound with a nitrogen atom bound to the carbonyl carbon.

amine. Compound with an **amino** or substituted amino group as its principal **functional group**.

amino. $-NH_2$ **group**.

amino acid. Molecule with the carboxyl **group** (–COOH) and the **amino** group ($-NH_2$), usually connected to the same carbon.

amorphous. Arrangement of **atoms** or **molecules** in which there is no regular pattern or order.

amphiphilic. Molecule that is part **polar** and part nonpolar.

amylase. Enzyme that **hydrolyzes** alpha 1→4 **glucose** links on **starch** chains.

amyloglucosidase. Enzyme that hydrolyzes all **glycosidic links** in **starch**.

amylopectin. Starch chain with branches every 30 to 50 units.

amylose. Unbranched **starch** chain comprising **glucose** connected by alpha 1→4 links.

anaerobic. Without air or oxygen.

anomeric. **Hemiacetal** or **hemiketal** carbon on a cyclic **carbohydrate molecule**. Has **bonds** to two oxygen **atoms**.

Archimedes principle. Law of physics: a body immersed in a **fluid** experiences an upward force equal to the weight of the fluid whose volume the body displaces.

aroma hops. Hops added shortly before the end of boiling to impart a characteristic aroma to the beer.

aromatic compound. **Organic compound** with ring containing alternating single and double **bonds**.

asymmetric carbon. Carbon **atom** with four different types of atoms or **groups (2)** attached.

atom. Tiny particle that is the smallest single unit of a particular **element**.

atomic number. Number of **protons** in an **atom**. The **atomic number** determines what the element is.

atomic weight. Molar mass of **atoms** of a particular **element**.

ATP. *See* **adenosine triphosphate**.

attenuation. Difference between initial and final solids content expressed as a fraction of the initial solids.

attract. Exert a force that pulls something closer.

awn. Long spiky extension of the **hull** of a seed of grain.

axial. Direction perpendicular to the plane of a ring.

Babylon. City in ancient **Mesopotamia**.

balanced. Having equal numbers of each type of **atom** on each side; said of a **chemical equation**.

bar. 1. A counter for serving alcoholic beverages. 2. An establishment that sells **alcoholic beverages** for consumption on the premises. 3. A unit of **pressure** equal to 100,000 pascals (0.987 atm).

barley. Grassy plant of the genus *Hordeum*.

barrel. 1. Large cylindrical container often made of wood. 2. Any of various measures of volume, for example, the U.S. beer barrel is 31 U.S. gallons (117.35 liters).

base. Substance whose **molecules** accept hydrogen **ions**.

basic. Dominated by **base**.

bauxite. Ore of aluminum with about 50% aluminum oxides and hydroxides and the rest mostly iron oxides.

beer. 1. Fermented **alcoholic beverage** made from a source of **starch** without concentrating the alcohol. 2. **Beer (1)** made with hops in contrast to **ale (1)**.

beer hall. A large establishment that sells **beer** for consumption on the premises.

beta-amylase (β-amylase). Enzyme that **hydrolyzes** 1→4 **glycosidic** bonds from the end of a starch chain to yield **maltose**.

binding site. Location on one **molecule** that has a specific attraction for another molecule or part of a molecule.

bittering hops. *See* **boiling hops.**

bitterness unit. Measure of bitterness by ultraviolet analysis, roughly equivalent to one milligram of isohumulone per liter. Also called an International Bitterness Unit.

bleach. Solution of about 5% sodium hypochlorite (NaOCl) in water. Used to remove color and as a disinfectant.

boiling hops. Hops added when boiling starts to add bitterness to beer.

bond. Force of **attraction** holding **atoms** together.

bond energy. Amount of **energy** needed to break a particular **bond**.

bound. Held to something by forces of **attraction**.

bouza. North African style of **beer** made from bread and **malt**.

breathalyzer. Portable device to measure **alcohol (2)** concentration in exhaled breath.

Brettanomyces. Genus of **yeast** used to make certain **beer** styles, notably Belgian lambic.

brewing liquor. Water that is to be made into **beer**.

brewhouse. The portion of a brewery with equipment for **mashing**, **wort** separation, boiling, wort chilling, and associated processes.

brewhouse efficiency. Ratio of potential **extract (3)** of carbohydrate sources in the mash to actual **carbohydrate** extracted.

brittle. Susceptible to cracking or shattering.

BTU. British Thermal Unit. Unit of **energy** equal to 1055 joules.

bung. Stopper for the side opening of a **cask**.

buoyancy. Force pushing a submerged or floating object upward.

calandria. Device to provide **heat** to a liquid to boil it.

calorie. Unit of **energy** equal to 4.184 joules.

Calorie. Unit of energy equal to 1000 **calories** or 4184 joules.

Campaign for Real Ale. CAMRA. Consumer movement to maintain the traditional forms of serving **ale (2)** to preserve a **cask**-conditioned flavor.

CAMRA. *See* **Campaign for Real Ale.**

carbohydrate. Substance with one **carbonyl group** and one –OH group on each of the other carbon **atoms**, or substance made by combining two or more of these.

carbonate. CO_3^{2-} **ion**.

carbonation. Dissolving carbon dioxide in water.

carbonyl. The C=O **group**.

carboxylic acid. Substance with the –COOH **group**.

carboy. Very large bottle made of thick **glass (1)** or plastic.

carrageenan. Polymer of sulfated galactose units prepared from seaweed. Used as a **fining** during the **wort** boil. Also called Irish moss.

cask. Barrel (1) with two or more openings used to transport and serve beer.

catabolite repression. Respiration enzymes are not produced when the **sugar** concentration is high. Also called glucose repression. *See also* **Crabtree effect**.

catalyst. Material that speeds up a chemical **reaction** but is not a product or a **reactant**.

cell membrane. Thin film surrounding a cell.

cellobiose. Disaccharide comprising two **glucose** units with a beta 1→4 link.

cellulose. Long chain **carbohydrate** comprising **glucose** units with beta 1→4 links.

centrifuge. Device to spin a sample rapidly causing heavier particles to separate from lighter ones by sedimentation.

channel. Protein structure extending through a **membrane** that can open to allow certain **ions** or **molecules** to flow through the membrane.

charcoal. Carbon made by heat treatment of plant or animal material.

chemical bond. *See* **bond**.

chemical equation. Symbolic statement of a chemical **reaction** showing how much of each item is consumed or produced.

chicha. Central and South American **beer** style made from **maize**.

chill haze. Cloudiness that appears at low temperature and disappears at room temperature.

chiral. Having a mirror image that is different from the original **molecule**.

chromatography. Method of separation and analysis that uses difference in affinities of the **molecules** under study for a stationary and a mobile phase.

coagulate. Form solid particles in a liquid.

coalescence. Particle growth by smaller particles colliding and sticking together to form larger particles.

Code of Hammurabi. List of laws from ancient **Babylon**.

coefficient. Number in a **chemical equation** indicating the number of **molecules** consumed or produced.

cohesive. Having **attractive** forces that hold parts of a substance together.

coke. Carbon made by heat treatment of coal.

cold break. Protein coagulation upon chilling of **wort**.

collagen. Protein in animal connective tissue.

column. Tube or cylinder.

coma. Deep state of unconsciousness.

component. One of the substances in a **mixture**.

composition. Relative amount of each **component** of a **mixture**.

compound. Single substance with more than one type of **atom**.

conditioning. Aging of **beer** to give improved clarity and flavor.

copper. 1. Highly conductive red **metal**, element 29, symbol Cu. 2. **kettle**.

core. Part of an **atom** including the **nucleus** plus **electrons** of lower **energy** than the outermost electrons.

corn. 1. Seed of grain. 2. **Maize**.

corrosion. Loss of **metal** due to **oxidation**.

countercurrent. **Fluid** being processed flows in the opposite direction of the fluid moving **heat** in or out.

covalent. Sharing **electrons**.

covalent bond. Chemical **bond** resulting from sharing **electrons**.

Crabtree effect. Alcoholic **fermentation** despite the presence of oxygen. Brought about a **sugar** concentration higher than the cell's capacity for **respiration**.

crown cap. Cap with puckers along its rim used for sealing bottles.

crystal. Regular three-dimensional pattern of **atoms**, ions, or **molecules** to form a solid.

crystalline structure. Arrangement of **atoms** or **molecules** into a regular pattern in three dimensions.

cuneiform. Earliest known writing system. Consists of markings made from wedges usually impressed in moist clay with a stylus.

decoction. Raising the temperature of **mash** by boiling a portion of it and returning the hot portion to the **mash tun**.

defect. 1. Property that fails to meet a quality standard. 2. Deviation from perfect regularity in the structure of a **crystal**.

degree of fermentation. Fraction or percentage of **original extract** that is converted during **fermentation**.

dehydration. Chemical **reaction** in which water is removed and a **bond** is formed between two **molecules** to make a larger molecule or between two parts of a molecule to make a ring.

deionize. Demineralize.

demineralize. Subject to **ion exchange** treatment in which undesired **ions** are replaced with hydrogen ions and hydroxide ions, which combine to yield water.

denature. Cause **protein molecules** to lose their shape and function as a result of high temperature, extreme **pH**, mechanical stress, and so on.

density. Ratio of mass to volume.

detector. Sensor for an analytical instrument.

detergent. Surfactant used for cleaning.

dextrin. Soluble **carbohydrate** with several **sugar** units resulting from partial **hydrolysis** of **starch**.

diacetyl. Ketone $CH_3C(=O)-C(=O)-CH_3$ that has a buttery flavor. Also called butanedione.

diastereomer. 1. One of a set of **isomers** distinguished by different configurations about **asymmetric carbons**, but that are not mirror images of one another. 2. One of a set of **isomers** with the **atoms** connected in the same way but directed differently in space that are not mirror images of one another.

diatom. Algae with a silicon dioxide-containing cell wall.

diatomaceous earth. Porous clay-like material composed of fossilized diatom cell walls, used as a filtration medium.

dimension. Type of measurable quantity, like length or speed.

dimethyl sulfide. $(CH_3)_2S$, gives an **off-flavor** to **beer**.

dipole–dipole interaction. Force of **attraction** between **polar molecules**.

disaccharide. Carbohydrate with two simple **sugar** units.

dispersion. Forces resulting from temporary **polarity**.

disproportionation. Particle growth by the dissolving or evaporating of material from small particles and the deposition of material onto larger particles. *See also* **coalescence**.

double bond. Covalent bond formed by sharing two pairs of elections.

draff. Spent grain after **mashing**.

EBC. *See* **European Brewing Convention**.

electrode. Solid electrical conductor that makes contact with a nonmetallic part of a circuit.

electron. Elementary negatively charged subatomic particle.

electronegativity. Tendency to attract shared **electrons** in a **covalent bond**.

electron group. Unshared pair of **electrons** or a **bond** (single, double, or triple).

element. Substance composed of **atoms** with a particular **atomic number**.

Embden–Meyerhoff pathway. Version of **glycolysis** used by all organisms higher than bacteria. Makes two **molecules** of pyruvic acid, two of **ATP**, and two of NADH for each **glucose molecule** consumed.

embryo. Baby plant or animal in a seed or egg.

emmer. Early form of wheat: *Triticum dicoccon*.

enantiomer. One of a pair of **compounds** that are mirror images of one another.

endopeptidase. Enzyme that **hydrolyzes protein molecules** in the interior of the **protein** chain.

endosperm. Part of a seed containing **starch** and some **protein**.

energy. Quantity measuring the capacity to make something move.

energy level. Value of the **energy** that a particle is permitted by physics to have. Particles may not have energies between the energy levels.

enol. **Compound** with an –OH group attached to a carbon **atom** that also has a **double bond** to another carbon atom.

enzyme. **Protein** that acts as a **catalyst** and speeds up specific chemical **reactions**.

equatorial. Direction in the plane of a ring pointing from the center to the outside.

ester. **Compound** formed by **condensation** of an **acid** and an **alcohol (1)** releasing water.

ethanol. Two-carbon **alcohol**, CH_3CH_2OH, also called ethyl alcohol.

ether. **Compound** with C–O–C as the principal **functional group**.

euphoria. Sense of extreme well-being.

European Brewing Convention. **Beer** standard setter for Europe. Also applied to any of their analytical methods, especially color. Abbrev. EBC.

exopeptidase. **Enzyme** that **hydrolyzes protein molecules** at points near the ends of the protein chains.

extract. 1. Soluble **carbohydrates** recovered from **mash**. 2. Fully or partly dehydrated **beer wort**. 3. Solids content of beer wort.

false bottom. Perforated plate to hold up grain during **wort** separation.

fermentability. Extent to which **carbohydrates** are usable by **yeast**.

fermentation. **Energy** production from a source of food without the consumption of oxygen.

fermentation, degree of. *See* **degree of fermentation**.

fermentation lock. Tube filled with liquid that allows gas out of the **fermenter** and prevents the entry of air.

fermented beverage. Liquid containing **ethanol** prepared for human consumption by **fermentation** of **sugars**.

fermenter. Vessel in which **fermentation** is carried out.

FG. **Final gravity**.

filtration. Purification of a **fluid** by passing it through a medium with small holes or channels.

final gravity. **Specific gravity** after **fermentation**.

fining. Material added to **beer** or **wort** to bind and precipitate small particles and clarify the beer.

flavor threshold. Lowest concentration at which a flavor **compound** can be detected by taste or aroma.

flavor unit. Concentration of a flavor-active substance given in multiples of the **flavor threshold**.

fluid. Material that flows. Liquid, gas, or **slurry**.

flux. Material added to lower a melting point.

foam. Particles of gas in a liquid or solid matrix.

forced carbonation. Adding carbon dioxide by introducing it under **pressure**.

formula. Representation of a **molecule** or **compound** using the symbols for the **elements** in it with subscripts to show how many **atoms** of each element are in a unit of the compound.

free radical. Atom, molecule, or **ion** with an unpaired **electron**.

freezing point depression. Lowering of the melting point of a substance by the presence of a dissolved material.

full mash. Process to make beer starting with preparation of wort by mashing grain in hot water.

functional group. Group of atoms that give a molecule characteristic behavior or properties.

fused quartz. Glass made from pure silicon dioxide. Used in high temperature applications.

fusel alcohol. Any **alcohol (1)** with three or more carbon **atoms** found in an **alcoholic beverage**.

gas chromatograph. Device to analyze **mixtures** of substances that can evaporate by allowing them to separate on a packed or coated **column**.

gelatin. Hydrolyzed form of **collagen** often derived from animal skin or bones. Used as a **fining**.

gelatinize. Cause the absorption of water between the chains of **molecules** of a substance made up of long chains, like **starches** or **proteins**.

geometric isomer. Set of **isomers** distinguished by the arrangement of **groups** on the same side or across from one another with reference to some feature of the **molecule**. An example would be the cis and trans forms of some **compounds** with double **bonds**.

geometry. *See* **underlying geometry**.

glass. 1. Brittle, transparent, noncrystalline mixture of silicon dioxide and other oxides. 2. Any noncrystalline solid that softens without a definite melting point as the temperature is increased.

glass electrode. Electrode with a special **glass** envelope making it selective for the hydrogen **ion**. Used in sensors of **pH** meters.

glucan. Any compound made by connecting **glucose** molecules with **glycosidic links**.

glucose. A sugar, $C_6H_{12}O_6$ in a specific form. Found in living cells.

gluten. 1. Mixture of storage **proteins** in seeds of grassy plants such as wheat, **barley**, and rye. 2. Protein or fragment that causes an adverse reaction in persons with celiac disease.

glycogen. Starch chain with branches every 10 to 12 **glucose** units. Used to store **energy** by animals and fungi.

glycolysis. Pathway to convert **glucose** to pyruvic acid. **ATP** and NADH are produced.

glycosidic link. Bond between an **anomeric** carbon atom on a **carbohydrate molecule** and an –OR **group**.

Godin Tepe. Prehistoric settlement on the ancient Silk Road in the Zagros Mountains in western Iran. Site of first chemical evidence of **barley beer**.

green beer. Beer before **conditioning**, also called ruh beer.

grist. Crushed grain.

group. 1. Column of the **periodic table**. 2. **Functional group**.

gruit. Mixture of herbs formerly used to flavor **beer**.

gushing. Premature or excessive release of dissolved gas from **beer**.

half-reaction. The **oxidation** portion or the **reduction** portion of an oxidation–reduction **reaction**.

Hall–Heroult process. Electrochemical method to make aluminum **metal** from aluminum oxide.

Hanse. Hanaseatic League.

Hanseatic League. Confederation of trading cities and guilds that operated along the North and Baltic Seas from 1159 until the 17th century.

hardness. High concentration of calcium, magnesium, or other **ions** with an electrical charge of +2 or more, said of water.

haze. Cloudiness.

head retention. Stability of foam on **beer**.

heat. Energy that flows from a region of high temperature to a region of lower temperature.

hedonistic. 1. Judgment based on liking or disliking. 2. Concerned with pursuit of pleasure.

hematite. Geological term for iron(III) oxide (Fe_2O_3).

hemiacetal. Product of the **reaction** of an **aldehyde** with an **alcohol**.

hemiketal. Product of the **reaction** of a **ketone** with an **alcohol**.

Henry's law. Gas dissolves in proportion to its **pressure** at the surface of a liquid.

heterogeneous. Having regions with different **compositions** or properties.

homogeneous. Displaying the same **composition** and properties in all regions down to the molecular level.

hop. Plant, *Humulus lupulus*, whose female flowers are used to flavor **beer**.

hop-back. Straining device using hop flowers as a filtering medium.

hopped wort. Wort after boiling with **hops**.

hops. Female flowers of the **hop** plant. Used to flavor beer.

hordein. Storage **protein** in **barley**.

Hordeum vulgare. Common **barley**.

hormone. Compound used by one part of an organism to signal other parts of the organism.

hot break. **Protein coagulation** during boiling.

huangjiu. Chinese **beer** style made from rice.

hull. Woody outer covering of a seed.

humulone. Principal **alpha acid** produced by **hop** flowers.

Humulus lupulus. Botanical name for the **hop**.

hydrate. 1. *n.* **Compound** whose **formula** includes one or more water **molecules**. 2. *vt.* React by the addition of water to a **compound**.

hydrated. Surrounded by water **molecules**.

hydrocarbon. **Compound** with only hydrogen and carbon.

hydrogen bond. Force of **attraction** between a hydrogen **atom** bound to an N, O, or F atom and an unshared pair of **electrons** on another **molecule** (or on a remote place on the same molecule).

hydrolysis. **Addition** of a water molecule across a **bond** in a larger **molecule** resulting in two smaller molecules.

hydrometer. Device to measure **density** of a liquid by how high the device floats in the liquid.

hydronium ion. H_3O^+ ion. Signature of an **acid** in water.

hydrophilic. **Polar**, charged, or **hydrogen bonding**, said of a **molecule** or a region of a molecule.

hydrophobic. Nonpolar, said of a **molecule** or a region of a molecule.

hydrophobic force. Tendency for **hydrophobic molecules** or regions of molecules to **repel** water, other highly **polar compounds**, or hydrophilic regions of molecules.

IKE. *See* **isomerized kettle extract**.

indicator. **Compound** whose color depends on **pH**.

induced fit. Mechanism (1) of **enzyme** function in which binding of the **reactant molecule** causes a change in the shape of the enzyme molecule.

intermolecular force. Force **attracting** electrically **neutral (1) molecules** toward one another.

ion. Electrically charged particle.

ion exchange. Method to soften or **purify** water by replacement of undesirable **ions** with acceptable ions.

ionic. Having or caused by charged particles.

ionic bond. Chemical **bond** resulting from force of **attraction** between a positive and a negative **ion**.

isinglass. **Collagen** derived from fish swim bladders. Used as a **fining**.

iso-alpha acid. Bitter isomerization product of an **alpha acid**.

isoelectric point. pH at which the electric charge on a molecule or other particle is, on average, neutral.

isohumulone. Bitter, soluble isomer of humulone formed during boiling.

isomer. One of a set of compounds with the same number of each type of atom and distinguished by differences in structure or geometry.

isomerization. Change in the form or structure of a **molecule** without adding or removing any **atoms**.

isomerized kettle extract. **Hop** extract that has been preisomerized intended for addition to the boiling kettle to adjust bitterness. Also called IKE.

keg. Metal cylinder for transporting and serving **beer**. Has one opening and a dip tube.

keg coupling. Valve to connect a **keg** to **beer** dispensing lines and a source of gas.

ketone. **Compound** with a **carbonyl group** attached to two carbon **atoms** (and no hydrogen).

ketose. **Monosaccharide** originating as a **ketone**.

kettle. Vessel for boiling **wort**.

keystone. Hole on the end of a **cask** to accept a valve.

kiln. Furnace or oven.

kvass. Eastern European **beer** made from bread.

lager. **Beer** fermented at temperatures below 60 °F.

lambic. Belgian style of **ale** made with naturally occurring **yeast**.

lauter tun. Vessel for **wort** separation using the spent grain as a filter medium.

Lewis structure. **Structural formula** for a **compound** showing **atoms** with their symbols and shared pairs of **valence electrons** as lines. Sometimes unshared valence electrons are shown as dots.

ligand. 1. **Ion** or **molecule** that provides an unshared pair of **electrons** forming a **covalent bond** to a **metal atom** forming a complex ion or coordination compound. 2. A small molecule that binds to a **protein**, usually by intermolecular (noncovalent) interaction.

lightstruck. **Off-flavor** in **beer** resulting from exposure to light. Also called "skunked."

linear. Forming a straight line.

lipid. Family of biological **compounds** whose **molecules** are at least partly **hydrophobic**.

lipid bilayer. Film-like structure comprising two layers of **molecules** with **polar** heads and nonpolar tails. The molecules in each layer are organized with nonpolar regions pointed to the other layer and polar regions directed away from the other layer. **Lipid** bilayers are the main structural element in **membranes**.

lipid raft. Region of a **membrane** with a high concentration of sterols. Serves as an anchor point for **proteins**.

liquefication. **Hydrolysis** of **starch** into water-soluble **molecules**.

Lovibond. Color scale used for **beer** and other foods.

lupulin. Yellow sticky powder produced by **hop** flowers. Source of **beer** flavor **compounds**.

Maillard reaction. Reaction starting with the formation of a Schiff base from a **sugar** and an **amino acid**. Can form large highly colored **melanoidin** molecules or smaller highly flavored molcules.

main group. Any of the eight columns of the **periodic table** comprising the first two and the last six columns.

maize. *Zea mays*, a grain called "corn" in the United States.

malt. 1. *n.* Sprouted and dried grain. 2. *v.* Make into **malt (1)**.

maltose. Sugar made of two **glucose** units connected by an alpha 1→4 link.

maltotriose. Sugar made of three **glucose** units connected by alpha 1→4 links.

mash filter. Device for **wort** separation by driving the wort through filters.

mash. 1. *v.* Initiate or cause **mashing**. 2. *n.* **Slurry** undergoing **mashing**.

mashing. Treatment of **malt** and other starchy materials with hot water to allow **enzymes** to **hydrolyze** the **starch** to **sugar**.

mash thickness. Ratio of volume of water to mass of grist in **mashing**.

mash tun. Vessel for **mashing**.

MBT. 3-Methylbut-2-ene-1-thiol. Gives lightstruck **beer** a skunky flavor.

mead. Alcoholic beverage made from fermented honey.

mechanism. 1. Sequence of elementary steps accounting for a chemical **reaction**. 2. Scientific doctrine holding that life processes can be accounted for by the normal laws of physics and chemistry. *See also* **vitalism**.

melanoidin. Highly colored **compounds** made up of large **polymers** produced from **sugars** and **amino acids** by the **Maillard reaction**.

membrane. 1. Thin film. 2. Thin film surrounding a cell or structure within a cell consisting of a **lipid bilayer** with attached **protein molecules**.

meniscus. 1. Curved surface made by a liquid climbing up a surface with which it is in contact. 2. Something with a crescent shape.

Mesopotamia. Region between the Tigris and Euphrates Rivers corresponding to parts of modern-day Iraq.

metal. Element with loosely held **electrons**. Typically has luster, has high electrical and thermal conductivity, and can often be bent or shaped without cracking. Metals often react by losing **electrons** to give positive **ions**.

metallic luster. Shininess in **metals**.

metalloid. Element with some properties characteristic of **metals** and some properties characteristic of nonmetals.

microbial filtration. Filtration to remove live microbes.

micropyle. Small opening in a seed or other structure.

mill. Device to crush or grind a material.

milling. Mechanically grinding or crushing a material.

mixture. A material with more than one substance.

modification. Changes that a grain undergoes during **malting**.

molar concentration. Number of **moles** of a **component** of a **solution** divided by the volume of the solution in liters.

molarity. **Molar concentration**.

molar mass. Mass of one **mole** of particular **atoms** or **molecules**.

mole. Amount of a substance containing 6.02205×10^{23} particles or **formula** units.

molecule. Group of **atoms** that are much more tightly bound to one another than to any other atoms. As a result they travel together as one distinguishable particle.

monastery. Religious community, usually of celibate men.

monk. Male member of a **monastery**.

monosaccharide. Simple **sugar**. **Aldehyde** or **ketone** with three or more carbon **atoms** and an OH **group** on all noncarbonyl carbons, or the **hemiacetal** or **hemiketal** resulting from internal **addition** of one of the OH groups across the C=O **bond**.

mouth feel. Component of flavor resulting from heat, cold, texture, or irritation of sensory cells in the mouth.

near infrared. Light whose wavelength is slightly longer than what can be seen with the eyes.

Neolithic era. New stone age, about 12,000 to 6700 years ago in the Middle East. Period in which agriculture was developed.

neuron. Nerve cell.

neurotransmitter. **Compound** that carries signals to or from **neurons**.

neutral. 1. Without net positive or negative electrical charge. 2. Neither **acidic** nor **basic**.

neutron. Electrically **neutral** elementary subatomic particle. Found in atomic **nucleus**.

noble gas. Any element in group 8A of the **periodic table**. Noble gases are chemically unreactive.

node. Joint in a plant.

nonmetal. **Element** that holds its **electrons** tightly. Nonmetals often react by gaining electrons to give negative **ions**, or by forming **covalent bonds**.

nucleation site. Particle or other feature that **attracts** dissolved **molecules** and allows them to come out of solution.

nucleus. 1. Positively charged particle at the center of an **atom**. 2. **Nucleation site**. 3. Membrane-enclosed region of a cell containing genetic material.

OE. *See* **original extract**.

off-flavor. Undesired flavor.

OG. *See* **original gravity**.

olfactory bulb. Part of the brain that processes aroma sensations.

oligosaccharide. A **molecule** with 3 to 10 **sugar** units connected by **glycosidic links**, usually without branches.

opaque beer. African family of **beer** styles often made with sorghum and consumed during active **fermentation**.

organic. Having carbon–carbon or carbon–hydrogen **bonds**.

organic chemistry. Behavior, synthesis, and study of **organic compounds**.

original extract. Mass percent solids in **wort** before **fermentation**.

original gravity. Specific gravity of **wort** before **fermentation**.

osmosis. The transfer of **molecules** through the pores of a **membrane** that can allow some molecules to pass but blocks others.

osmotic pressure. The **pressure** that stops flow through a **semipermeable membrane**.

oxidation. Chemical **reaction** in which **electrons** are lost.

oxidation number. Fictitious charge determined by allocating all **electrons** shared by two **atoms** to the more **electronegative** of the pair. If the atoms are of the same **element**, the shared electrons are divided evenly.

packing. Powder used as a stationary phase in **chromatography**.

partial pressure. The **pressure** that a component of a gas **mixture** would exert if it were the only gas in the container.

pasteurization. Preservation of foods by brief heating to a moderate temperature (typically 63 °C = 145 °F) to kill microbes.

peptidase. Enzyme that breaks the chains of **protein molecules**.

peptide. Polymers of 50 or fewer **amino acids**.

period. Row of the **periodic table**.

periodic table. Chart of chemical **elements** in order of **atomic number** and organized in columns by the number of **valence electrons**.

permanent hardness. Hardness in water that is not removed by boiling.

permeation. Penetration of a solid barrier by a gas or liquid.

peroxide. Compound with –O–O– **functional group**.

pH. Measure of the **acidity** or **basicity** of a **solution**. More acidic gives a lower pH, more basic gives a higher pH.

pH electrode. *See* **glass electrode**.

pH meter. Electrochemical device to measure **pH** in water **solutions**.

pH paper. Paper impregnated with **pH indicators**. Compared to a color scale to estimate pH.

phase. Region of a sample with a particular **composition** and properties.

phenol. 1. Compound: C_6H_5OH. 2. Family of **compounds** with an –OH **group** directly attached to an **aromatic** ring.

photon. Smallest unit of light.

Pietism. Christian religious movement influential in Protestant thought and in the founding of Methodism.

PIKE. Isomerized hop extract that has been neutralized with potassium hydroxide (KOH) to improve solubility.

Pilsen. Town in Bohemia, in what is now part of The Czech Republic. Home of the Pilsener **beer** style.

pipe scale. Hard deposits adhering to the insides of pipes.

pitch. Add **yeast** to **ferment beer**.

plasma membrane. *See* **cell membrane**.

point. Measure of specific gravity equal to $1000 \times (SG - 1)$. For example, a specific gravity of 1.044 is equivalent to 44 points. Four points is equivalent to about 1% dissolved **sugar**.

polar. Having a positively and a negatively charged end.

pollutant. Undesired **component** of a **mixture**.

polyatomic ion. Collection of two or more **covalently bound atoms** with an electric charge (e.g., **carbonate ion**).

polymer. Large **molecule** made of subunits connected by **covalent bonds**.

polysaccharide. Carbohydrate consisting of many **monosaccharide** units connected with **glycosidic links**.

polyvinylpolypyrrolidone. PVPP. **Polymer** used as a **fining**. Cross-linked, insoluble version of polyvinylpyrrolidone.

precipitate. 1. To come out of **solution**, especially as a solid. 2. The solid that comes out of solution.

pressure. Force divided by area.

primary structure. Sequence of **amino acids** in a **protein**.

product. New material made by a chemical **reaction**.

prohibition. Laws against manufacture, sale, or use of **alcoholic beverages** or other substances. Specifically, national prohibition on **alcohol (2)** in the United States from 1919 to 1933.

protein. Polymer of 50 or more **amino acids**.

proton. Positively charged elementary subatomic particle. Found in atomic **nucleus**.

psychoactive. Substance that alters brain function.

pub. *See* **bar (2)**.

pump. 1. Device to drive a **fluid** from place to place. 2. **Protein** structure extending through a **membrane** to drive specific **ions** or **molecules** from regions of lower concentration to regions of higher concentration.

purification. Removal of undesired **components** from a **mixture**.

PVPP. *See* **polyvinylpolypyrrolidone**.

pyruvic acid. $CH_3COCOOH$, an intermediate **compound** in **respiration** and **fermentation**.

reactant. Material that is consumed by a chemical **reaction**.

reaction. Process in which one or more substances, the **reactant(s)**, are converted to one or more other substances, the **product(s)**.

reactive oxygen species. Molecules, **ions**, or **free radicals** containing oxygen in a form that reacts more easily than oxygen molecules.

real extract. Mass percent solids in decarbonated **beer** after **fermentation**. Abbrev. RE.

receptor. Structure that selectively binds certain **molecules** or **ions** and causes a response inside a cell.

redox. Class of chemical **reactions** involving an **oxidation** and a **reduction**.

reducing sugar. **Sugar** with a free OH **group** (not bound by a **glycosidic link**) on the **anomeric** carbon.

reduction. Chemical **reaction** in which **electrons** are gained.

reference electrode. Electrode kept in electrical contact with a sample to complete the circuit for electrochemical measurements like **pH**.

refractive index. Ratio of the speed of light in vacuum to that in a particular sample.

refractometer. Instrument to measure **refractive index**. Can be used to determine **sugar** content in **beer wort** and fruit juice.

Reinheitsgebot. Bavarian law restricting **beer** ingredients to water, **barley**, and **hops**.

repel. Exert a force driving something away.

resonance. Delocalization of **electrons** represented as the average of alternative forms of a **molecule** differing only by the placement of the **valence** electrons.

respiration. Conversion of air and food to carbon dioxide and water. Releases **energy**.

reverse osmosis. Method of purification by forcing a liquid through a semipermeable membrane from the impure side to the pure side.

ribes. Catty flavor in **beer**.

rotation. Motion of a **molecule** involving turning about an axis.

ruh beer. *See* **green beer**.

saccharification. **Hydrolysis** of **sugar** units from the end of a **starch** chain to give a sugar **molecule**.

Saccharimyces cerevisiae. Top fermenting **yeast** used to make ale and bread.

Saccharimyces pastorianus. Bottom fermenting yeast used to make lager beer.

sake. Unhopped rice beer from Japan. Also called rice wine.

saloon. **Bar (2)**, usually with tables and chairs.

sanitize. Treat to reduce number of microbes to an acceptable level.

sanitizer. Substance used to sanitize surfaces.

scutellum. Separator between the **endosperm** and **embryo** compartment in a seed.
sediment. 1. *v.* Fall to the bottom of a container. 2. *n.* Material that has fallen to the bottom of a container.
seed coat. Waxy covering for a seed beneath the **hull**.
semimetal. *See* **metalloid**.
semipermeable membrane. A **membrane** through which certain molecules or **ions** can pass, but not others.
shape. Spacial arrangement of **atoms** in a **molecule** not including unshared electron pairs.
shell. Allowed **energy level** for **electrons**. Different shells can accommodate different numbers of electrons.
silica gel. Porous silicon dioxide with bound water. Used as a **fining**.
siphon. Tube to transfer a liquid over a barrier from a higher to a lower level.
six-row barley. Variety of **barley** in which all three flowers in each group are fertile.
skeletal structure. Representation of a **molecule** in which carbon atoms are shown as the ends of line segments and the hydrogen atoms and the bonds to them attached to carbon atoms are omitted.
slurry. Mixture with solid particles suspended in a liquid.
small pack. Can or bottle sold for consumer use at home.
soda lime glass. **Glass** prepared from sand with sodium carbonate and calcium carbonate used as a **flux**.
soft water. Water with low concentrations of calcium and magnesium **ions**. *See also* **hardness**.
solute. Minor **component** of a solution.
solution. Mixture with the same **composition** throughout.
solvent. Component that makes up the majority of a **solution**.
sparge. Wash **wort** from a bed of grain with hot water.
spear. Tube that goes from the opening to the bottom of a **keg**.
specific gravity. Mass of the volume of a material that has the same volume of one mass unit of water. Equivalently, the ratio of the **density** of a sample to the density of water. Used as a measure of the **concentration** of dissolved **carbohydrates** in **wort** and beer.
specificity. Property of an **enzyme** (or other **catalyst**) to cause a **reaction** in a certain **compound** more than in other potential reactants.
spigot. Valve for liquid.
spile. Peg driven into the **bung** of a **cask** to control the internal **pressure**.
SRM. *See* **Standard Reference Method**.
stacking force. Force of **attraction** between flat, ring-shaped **molecules**.

Standard Reference Method. Standard method for **beer** color of the American Society of Brewing Chemists (ASBC). Called SRM. Officially *ASBC Methods of Analysis*, Beer 10.

starch. Long chain of **sugar molecules** in a form that is readily **hydrolyzed**. Usually the links are mainly alpha 1→4.

starter. Culture of **yeast** grown in a small batch of **wort** used to initiate **fermentation**.

stereoisomer. One of a set of **molecules** with the same **atoms** bonded in the same order, but differing in the way they are directed in space.

sterilization. Treatment to kill or remove all microbes and spores in a material or on an object.

sterol. Member of a family of **alcohols** in which the –OH **group** is bound to a **molecular** plate consisting of four fused rings.

strike water. Water added to **grist** to begin **mashing**.

structural formula. Representation of a **molecule** showing the **atoms** and the **covalent bonds** connecting them.

Student's t-test. Statistical method to determine significance without knowledge of the standard deviation of the population.

stupor. Lowered level of consciousness and response.

substituent. Atom or **functional group** that replaces a hydrogen atom in an **organic compound**. The substituent is of lower priority than the principal **functional group** of the **molecule**.

sugar. Monosaccharide or **disaccharide**.

Sumer. Ancient civilization that existed 7000 to 4000 years ago on the lower reaches of the Tigris and Euphrates Rivers.

surface. Boundary where **phases** meet.

surface energy. Energy to overcome cohesive forces and make new **surface**.

surface tension. Amount of **surface energy** needed to make one unit of **surface**.

surfactant. Substance that selectively migrates to a **surface**.

tap. 1. Spigot or fitting to take **beer** from a **cask** or **keg**. 2. To install such a fitting.

tare. Weight of a container; must be subtracted from the total weight to get the weight of the contents.

taste bud. Organ containing cells responsible for the sense of taste.

tavern. Shop for purchase and consumption of **alcoholic beverages** and usually food on the premises. Some also provide lodging.

temperance movement. Social movement urging reduced consumption of **alcoholic beverages**. Followers advocated diverse programs ranging from voluntary moderation of consumption to enforced **prohibition**.

temporary hardness. Water **hardness** that is removed by boiling, indicating that the negative **ion** is **carbonate**.

terpene. Any member of a family of naturally occurring **hydrocarbons** with one or more double **bonds**.

tetrahedral. Three-dimensional **geometry** with four **electron groups** arranged at roughly 109.5 degree angles about a central **atom**.

tautomerism. Shift of a hydrogen **atom** resulting in the exchange of a double **bond** with an adjacent single bond.

thermometer. Device to measure temperature.

thiol. Compound whose principal **functional group** is the –SH group.

threshold potential. Electrical potential in a cell that causes **channels** to admit positive **ions** generating an electrical signal.

toast. Benediction recited over an **alcoholic beverage**.

Tollens's reagent. **Basic** solution of $Ag(NH_3)_2^+$ in water used to test for reducing **sugars**.

top fermenting. **Yeast** that rises to the top of the **fermenter** during **fermentation**. This is due to a slightly **hydrophobic** surface.

tough. Resistant to cracking and shattering.

transduction. Process in which a stimulus to a cell causes a signal that can be transmitted to another cell.

transition state. Position in the progress of a chemical **reaction** encounter at which the potential **energy** is highest.

translation. Motion of **molecules** in which the whole molecule moves from place to place.

trans-2-nonenal. Compound $CH_3(CH_2)_5CH=CHCHO$ that gives a cardboard flavor to stale **beer**.

trigonal planar. Geometry with three **electron groups** all in the same plane arranged at roughly 120 degree angles about a central **atom**.

triple bond. **Covalent bond** formed by sharing three pairs of **electrons**.

trub. Sediment from **wort** or **beer** that **precipitates** at various stages of brewing. It could include **coagulated proteins**, bits of **hops**, inactive **yeast**, and fatty solids.

two-row barley. Variety of **barley** in which only the central flower in each group of three is fertile.

underlying geometry. Spatial arrangement of **electron groups**, including the unshared **electron** pairs, in a molecule.

unit. Standard quantity of measure, like a kilogram (unit of mass).

valence. Outermost **electrons** in the highest occupied **shell** of an **atom**.

valency. Normal number of **covalent bonds** that an **atom** forms in neutral **molecules**.

vapor. Gas at a low enough temperature that it can be liquefied.

vapor pressure. **Pressure** at which a vapor is in equilibrium with its liquid at a particular temperature.

vibration. Motion of **molecules** in which **bonds** bend or stretch.

viscosity. Resistance to flow in a **fluid**.

vitalism. Scientific doctrine that holds that processes in living organisms cannot be explained solely by the laws of chemistry and physics. Today, vitalism is rejected by mainstream science.

volatile. Easily evaporated or boiled.

Volstead Act. U.S. law adopted in 1919 providing for the enforcement of national **prohibition** of **alcohol (2)**.

voltage gated channel. **Channel** in a cell **membrane** that opens in response to the electrical potential being less negative than the **threshold potential** allowing particular positive **ions** to cross the membrane.

vorlauf. Recirculate **wort** from the **lauter** or **mash tun** through the bed of grain to clarify the **wort**.

water softener. **Ion exchange** device that replaces undesired ions with sodium and chloride ions.

whirlpool. Tank in which beer **wort** is made to flow in a circle, causing suspended solids to settle in the center.

wort. **Sugar** solution extracted from grain by mashing.

wort chiller. Device to remove **heat** from **wort** after boiling.

yeast. Single celled fungus, certain species of which are used to make **beer**, wine, and bread.

zwickel. Spigot in a **fermenter** for sampling the **beer**.

zymurgy. Chemistry of **fermentation**.

INDEX

Absolute configuration   117
Absorbance   187
Acetic acid   73, 172–174, 190
Acetyl coenzyme A   174
Acid   43, 71, 73, 103
  hydroacid   43
  hydrochloric   71
  name   43, 103
  oxyacid   43
  strong and weak   72
Adalhard of Corbie   5
Addition reaction   95, 100, 113, 155
Adenosine triphosphate (ATP)   165–171, 199–200
Adjunct   17, 144
Alcohol   10, 92, 96, 103, 113, 171. *See also* Ethanol
  blood   190
  calories   146
  content   187
    by near infrared   188
    by specific gravity   188
  dehydration   97
  formula   96
  fusel   172
  higher   172
  hydrometer   187
  naming   97
  sake   9
  temperature   172
  by volume   178, 187, 189
Aldehyde   100, 113, 126, 252
  naming   102
Aldose   113
Ale   6, 265
  bitter style   212
  malty style   212
  phenolic style   213
Alkaloid   99
Alkane   92
  hydrophobic   92
  naming   93
Alkene   94
  naming   96
  substituent   96
Alkyl group   92, 93
Alkyne   95
  naming   96

The Chemistry of Beer: The Science in the Suds, First Edition. Roger Barth.
© 2013 John Wiley & Sons, Inc. Published 2013 by John Wiley & Sons, Inc.

Alpha acid 153, 155, 220, 221
  isomerization 219
  isomerized 153
Aluminum 241, 245
  alumina 244
  can(s) 246
  cost 246
  electrolysis 246
  Hall–Heroult process 246
  history 245
  reduction 246
American Society of Brewing Chemists 186
Amide 106
  base 106
Amine 98
  amino acid 135, 137–140
  base 99
  naming 100
  primary 98
  secondary 98
  tertiary 98
Amino acid 135, 137–140
  enantiomer 137
  R group 137
Amino group 98
Ammonia 42, 72
Amphiphilic 236
Amylase 25, 134, 141, 144, 146
  alpha 134, 220
  beta 134, 220
  mechanism 141
Amyloglucosidase 146
Amylopectin 23, 122
Amylose 23, 121
Anomeric carbon 115, 120, 126
  configuration 119
Archimedes Principle 178
Aroma 195, 196, 200
  receptor 198
  transduction 199
Aromatic compound 94
  stability 95
Aromatic ring
  DNA 95
  phenol 97
Aspartame 197
Assyria 3

Atom(s) 33–35
Atomic weight 60. *See also* Molar mass
ATP, *see* Adenosine triphosphate
Attenuation 182
Axial position 119–120

Babylon 3
Barley 3, 18–19, 125, 131
Base 71
  strong and weak 72
Bavaria 7
Beer. *See also* Beer styles
  African 8
  ale 173, 265
  American 10
  appearance 23, 225, 236
  bitterness 198
  bouza 8
  from bread 3
  bright 228
  calories 146
  Calories 21
  carbohydrate 22
  carbonation 30, 231, 235, 237, 288
  cauim 9
  cheongiu 9
  Chibuku 8
  chicha 8
  clarity 73, 225, 227
  color 216, 217
  commerce 7
  composition 17
  consumption 7
  definition 17
  economics 12
  Egypt 4
  ethanol content 10
  filtration 228, 230
  flavor 3, 17, 23, 225
  as food 21
  gluten 22
  goddess 3
  gushing 27
  haze 49, 53
  head 225. *See also* Foam
  and health 21
  heavy 178
  history 1, 227

huangjiu 9
ingredients 8, 17
ions 76
kvass 8
lager 173, 265, 289
lambic 204
light 146
light absorption 186
lightstruck 154
lipids 257
low alcohol 146
malt liquor 146
masato 9
medicine 3
opaque beer 8
origin 1
oxygen 257
package filling 247–248
packaging 30
pH 79
Pilsner lager 7
prestige 12
protein 229
ration 5
recipe 216, 290
regulation 10
sake 8, 9
skunky 272
social role 8, 12, 13
sorghum 8
sour 204
South American 8
stale 251, 257, 260
standard American 7
status 12
storage temperature 260
style 7, 9, 237
taxation and regulation 6, 7
testing 177
water content 69
wheat 18
Beer image 12
Beer styles
 American lager 214
 bitter ale 212
 bitter lager 214
 Bock 214, 215, 219–221
 characteristics 211, 215
 classification 211
 families 211
 Kolsch 214
 lambic 214
 malty ale 212
 Munich dunkel 214
 Munich helles 214
 origin 211
 original gravity 215
 Pilsner lager 214
 realizing 215
 specialty 214
Beer's law 187
Benedict of Nursia 5
Benedict's reagent 126
Beowulf 4
Bicarbonate ion 76, 79
Bisphenol A 247
Bitterness 219. *See also* International Bitterness Unit
 calculation 219
 fermentation loss 219
 specific gravity 219
Boil/boiling 152, 279
 dimethylsulfide 205
 function 25
 hops 25
 hot break 25
 S-methylmethionine 205
Bond 36
 angle 44, 46, 50, 89
 covalent 37, 40, 89
 glycosidic 131
 hydrogen 48, 50
 ionic 38
 metallic 244
 multiple 41, 43, 95, 163
 polar 46
 rotation 92
Bottle 243
 crown cap 248
 image 247
Bottling 286, 287
Breathalyzer 190
Brettanomyces 21
Brewhouse
 capacity 151
 efficiency 217

Brewing
  ale 29
  boiling 25
  chilling 27
  conditioning 28–29
  cost 265, 285, 290
  Egypt 4
  extract 265, 283, 284
  filtration 29
  fire 7
  full mash 267
  history 225
  home 265
  lager 29, 289
  lauter process 25
  mash filtration 25
  mashing 23
  milling 23
  partial mash 265
  pot 268
  recipe 290
  safety 266
  sake 9
  sparging 25
  starter systems 288
  wort separation 25
Brewing liquor 69
British breweries 149
Bubble 231, 235
  nitrogen 237
Buchner, Eduard 10
Buoyancy 178, 235
Burton upon Trent 218

Cagniard-Latour, Charles 9
Cahn–Ingold–Prelog system 117
Calandria 152
Calcium 38
Calcium carbonate 243
Calcium oxalate 1
Calories, food 22
cAMP 199
Campaign for Real Ale 242
Can(s) 246
  advantage 247
  coating 247
  decoration 246
  filling 246
  image 247
  lining 246
  manufacture 246
  top 246
Carbohydrate 113, 123, 131, 134. *See also* Carbohydrate content
  elements 113
  recognition 123
  specific gravity 215
  test 126
Carbohydrate content 215
  attenuation 182
  degree of fermentation 182
  original extract 182
  real extract 182
  refractive index 182
  specific gravity 177
Carbon, asymmetric 115, 137
Carbonate ion 76
Carbonation, volumes 220
Carbon dioxide 76
  flavor 237
  mouth feel 220
  shape 43
Carbonyl 100
Carbonyl compound 257
Carbonyl group 126
Carboxylic acid 103, 106, 142, 172
  amino acid 137
  naming 103–104
Carrageenan 229
Cask 29, 30, 241
  bung 242
  keystone 241
  spile 242
  wood 241
Catabolite repression 171
Catalyst 23, 135
  enzyme 131, 135
Celiac disease 22
Cell(s) 161–163, 195–196
Cellobiose 120, 121, 126
Cellulose 121, 126
Centrifuge 29
Cgs system 58
Chain, stability 41
Channel 195, 197
  calcium ion 200
  voltage gated 200
Chemical equation(s) 52

Chemical reaction, food   21
Chemistry
  acid-base   71, 190
  atomic theory   33
  beer style   211
  biochemistry   134, 138, 165
  carbohydrate   113
  chemical formula   89
  compound   36, 42
  history   9
  hop products   154
  measurement   56
  metric system   57
  molecule   40
  organic   89
  protein   135, 137, 140
  reaction   52, 135, 137
  water   69
Chill/chilling   158, 205, 280, 284
Chiller   270
Chirality   115
Chlorine   38
Coagulate   157
Coalescence   227
Cold break   158
Collagen   230
Color   185, 198
  adjusting   219
  boiling   219
  melanoidin   216
Compound   36
  formation   37
  ionic   42
  mass relationship   63
  molar mass   62
  name   42
  nonmetal   42
  number prefix   42
Condensation reaction   120
Conditioning
  cask   242
  diacetyl   222
  vicinal diketone   222
Copper 152. *See also* Kettle
Countercurrent   158
Crabtree effect   171
Craft breweries   149
Criss-cross rule   39
Cryolite   246

Crystal   38, 245
  defect   245
Cuneiform   1
Cysteine   137, 138
Czech Republic   7

Deionizer, regeneration   83
Demineralizer, regeneration   83
Density   177
Detergent   227
Dextrin   23, 25, 146
  hydrolysis   146
  mouth feel   220
Diacetyl   222
Diastereomer   118
Diatomaceous earth   29
Dicarboxylic acid   104
Dimethyl sulfide   153
Disaccharide   120
Disproportionation   227, 237
Dublin   218

Edinburgh   218
EDTA, *see* Ethylenediamine tetraacetate
Efficiency, brewhouse   217
Egypt   4
18th Amendment   11
Electrode
  glass   191
  reference   191
Electron   244
  allocation   253
  cloud   33
  energy level   34
  gain and loss   253
  group   43
  shared   33, 37, 40
  shell   34
  transfer   33, 252
Electronegativity   46, 253
  table   47
Element(s)   33. *See also* Periodic table
  atomic number   33, 34
  vs. ion   38
  molar mass   34, 60
  name   34
  symbol   34
Embden–Meyerhof pathway   168
Embryo   131

Emmer 3
Enantiomers 116
Energy 36
  activation 135
  ATP 165, 167
  bond 165
  coal 212
  currency 165
  fermentation 27–28, 161, 170
  from food 21
  heat 158
  level 34
  resonance 167
  respiration 171
  surface 225, 227
  unit 21
Enol 101
Enzyme 8, 23, 131, 165
  active site 141
  amylase 25, 141, 146
  amyloglucosidase 146
  binding site 138
  catalyst 138
  denature 25, 141
  induced fit 138
  light beer 146
  malt 19
  malt liquor 146
  mechanism 141
  peptidase 145
  specificity 138, 141
Epichlorohydrin 247
Epoxy 247
Equatorial position 119–120
Equilibrium 72
Ergosterol 162
Erythrose 118
Ester 104, 173, 202, 212
  acyl part 105
  ale and lager 174
  alkyl part 105
  naming 105
  temperature 174
Ethanol 17, 28, 96, 161, 189
  addiction 10
  flavor threshold 201
  psychoactive 10
  regulation 10, 12
  synthesis 170
  tradition 13
Ether 98
  cyclic 98
  naming 98
Ethylenediamine tetraacetate (EDTA) 77, 78
Euphrates 3
Europe 4
European Brewing Convention 186
European Union 7
Extract 134
  original 182
  real 182
  temperature 144

False bottom 149, 150, 269
Fehling's reagent 126
Fermentability 134
  temperature 144
Fermentation 9, 27–28, 113, 161–174
  degree 182, 189
  energy 170
  heat 28
  lock 273, 281
  products 27–28, 204, 212, 214, 222
  reaction 27–28, 161
  secondary 282
Fermenter, bucket 271
Final gravity 178
Finings 228, 229, 242
  carrageenan 229
  collagen 230
  isoelectric point 229
  PVPP 230
Flavor 195, 196, 200
  alcohol 172
  background 203
  cardboard 251, 252
  catty 251, 252
  diacetyl 204
  dimethylsulfide 205
  ester 172, 173, 221
  fermentation effect 28
  fermentation temperature 221
  ferulic acid 213
  fruit 221
  hops 20
  isoamyl acetate 172

lightstruck 206
malt 19
malty 204, 220
MBT 206
microbe reduction 30, 249
nonenal 205
off-flavor 153, 157, 158, 171, 204
oxygen 252
phenolic 204
primary compounds 202
ribes 251. *See also* Flavor, catty
secondary 202
skunky 206
smoke 204
specialty 204
stability 204, 205, 251
tertiary 203
threshold 201
trans-2-nonenal 205. *See also* Flavor, nonenal
transduction 196
unit 201, 214
4-vinylguaiacol 213
Foam 27, 73, 225, 231
  collapse 231
  issues 236
  oil 236
  stability 231, 236
Force
  dipole–dipole 49
  dispersion 48
  electrical 34
  hydrogen bond 48, 50, 70
  hydrophobic 50
  intermolecular 48
  stacking 48
Formula 39
  structural 40, 89
  subscript 39
Free radical 207, 255
  abstraction 256
  addition 256
  instability 255
  interconversion 257
  resonance 255
  trap 261
Freezing point depression 243
Fructose 126

Functional group 91, 92, 137
  guide 107
  humulone 153
  multiple 92
Fusarium 237

Galactose 112, 229
Gas 231
  Boyle's law 233
  constant 232
  dissolved 231
  Henry's law 220
  ideal gas law 232
  mixture 233
  partial pressure 233
  pressure 232
  solubility 220
Gas chromatograph 187
Gelatin 230
Geometry
  bent shape 44, 45
  linear 43
  molecule 43
  tetrahedral 44
  trigonal planar 43
  tripod shape 45
Gildas 5
Glass 243, 272
  amorphous 243
  blowing 243
  bottle 30
  brittle 243
  brown 243
  flux 243
  fused quartz 243
  soda lime 243
  structure 243
Glucan 120, 124–125
Glucose 115, 119, 121, 126, 131, 161
  diastereoisomers 119
Glycogen 124
Glycolysis 168, 170
Glycosidic link 120
  alpha and beta 120
  hydrolysis 135, 146
  in polysaccharide 121
  reducing sugar 126
Godin Tepe 1
Gossett, Willian Sealy 10

Grist   132, 145, 149, 277
Gruit   5, 6
Gum   124–125
Gushing   237

Half-reaction   244
Hammer mill   132
Hammurabi   3
Hanseatic League   5
Hansen, Emil Christian   10
Hardness   77
    conversion   78
    measurement   77
    temporary   84
    units   78
Haze   227
    chill haze   228
    measuring   229
Head. *See also* Foam
    retention   236
    measurement   236
Hemiacetal   100, 114, 258
Hemiketal   101, 114
Henry's Law   231, 235
Hexahydro-isohumulone   155
Higher alcohol synthesis   172
Hop-back   157
Hops   5, 17, 153, 276
    alpha acid   219
    aroma   153, 221, 279
    beer preservative   6
    bitter   20
    boiling   279
    cultivation   5, 20
    dry hopping   153, 221
    extract   155
    Hallertau   220
    history   5
    hydrogenated extract   155, 207
    isohumulone   101
    isomerized extract   155
    isomerized pellets   155
    late hopping   221
    pellets   154, 155
    plant   20
    processing   21
    utilization   219
    varieties   20, 221
Hormone   20

Hot break   157
Hull   124, 125, 149
Humulone   153
Hydrate   84
Hydrogen, parked on NAD   170
Hydrogen bond, PVPP   230
Hydrogen bonding   155
Hydrogen ion
    proton   71
    transfer   71
Hydrogen peroxide   257
Hydrolysis   135, 136
    at branch   146
    enzyme   131
Hydrometer   178, 271
    reading   179
    scale   181
    temperature correction   180
Hydronium ion   72
    equilibrium   72, 73
    pH   73
Hydroperoxyl radical   256, 257
    reaction   256, 257
Hydrophilic compounds   51, 227, 236
Hydrophobic compounds   51, 227, 236
Hydroxide ion   255
    base   72
    equilibrium   72, 73
Hydroxyl radical   257
Hymn to Ninkasi   3

IBU, *see* International Bitterness Unit
Indicator   190
International Bitterness Unit   153, 201
International System   57
Iodine test   126
Ion   37, 38
    formation   37
    formula   38
    hydration   70
    hydronium   71
    hydroxide   72
    metal   69
    polyatomic   39
Iron, in hematite   244
Isinglass   230
Iso-alpha acid   155, 198, 206, 207, 219

Isoelectric point 75, 229
  collagen 230
  silica gel 230
Isohumulone 101, 153
  enol 101
Isomer 89
  cis-trans 95
  geometric 95

Keg 30, 241
  carbon dioxide 243
  coupling 242
  microbe reduction 243
  size 242
  spear 242
Keto-enol tautomerism 101
Ketone 100, 126
  naming 102
Ketose 113
Kettle 25, 149, 152
  design 157
Kjeldahl, Johan 10
Koji 9
Kosher laws 230
Krebs cycle 170

Lager 6
  bitter style 214
  Bock 215
  homebrewing 289
  malty style 214
  yeast 276
Lauter tun 125, 150
  knives 150
Le Chatelier's principle 72, 76, 83
Lewis, G. N. 38
Lewis dot diagram 38
Lewis structure 89
Liebig, Justus von 9
Ligand 261
Light 185, 198
  absorption 186
  scattering 227, 228
  wave 185
Lightstruck beer 30, 154, 201, 207
  inhibiting 155, 207
Linoleic acid
  in barley 257
  oxidation 257, 258

Lipid
  bilayer 162, 195
  raft 162
  removal 260
  trub 260
Liquification, alpha amylase 134
Lovibond 186
Lupulin 153

Maillard reaction 212, 216
  boiling 220
  color 212, 213
  conditions 213, 216
  mashing 220
Maize 8
Malt 3, 17, 274
  barley 18, 131
  black patent 216
  caramel 213. *See also* Malt, crystal
  color 213, 216
  color determination 216
  color potential 217
  color unit 219
  crystal 213, 216, 220
  data 216
  definition 18
  extract 283
  grades 20
  kilning 216
  modificatiion 19
  Munich 220
  pale 212
  Pilsner 216
  potential extract 217
  roasted 213, 216
  smoke 212
  stewing 213
Maltose 120, 126
Manioc 9
Mash 277
  filter 150
  filtration 132
  pH 79
  set 125
  temperature 145, 277
  thickness 145
  tun 145, 150, 268

Mash in, temperature 277
Mashing 25, 113, 131, 133, 149
  amylase 141
  decoction 144, 220
  gelatinization 23, 133
  liquefication 23, 134, 141
  saccharification 23, 134, 141
  temperature 141
Mass percent 64
MBT
  light 207
  threshold 201, 207
Mead 5
Measurement
  alcohol 187
  carbohydrate content 182
  color 186
  dimensions 57
  International System 57
  metric 57
  pH 190
  refractive index 182
  specific gravity 10, 178
  temperature 10, 183
  units 57
Mechanism doctrine 10
Melanoidin(s) 212, 216
Membrane 195
  channel 195
  fluidity 171
  pump 195
  receptor 196
Meniscus 179
Mesopotamia 1
Metal 244. See also Aluminum;
    Copper; Stainless steel
  corrosion 244
  crystalline structure 245
  electrical conductivity 245
  ions 257
  luster 245
  properties 36, 245
  reduction 244
  thermal conductivity 245
  toughness 245
  valence electron 244
Metric prefix 58
Meura Company 151

Microbe reduction 30, 249
Microbial filtration 249
Mill 132
  gap 132
Milling 131
  dust 132
  particle size 132, 149
  wet 133
Mirror image (chirality) 115
Mixture 36, 53, 233
  component 36, 53
  composition 53, 64
  heterogeneous 53
  homogeneous 53
  phase 53
Molar concentration 65, 78
Molarity 65. See also Molar
    concentration
Molar mass 60
  calculating 62
Mole
  calculating 63
  equations 63
Molecular motion 228
Molecule 40
  diatomic 36
  kinetics 51
  motion 51, 231
  polar 47, 49
  rotation 51
  shape 43
  temperature 52
  translation 51
  underlying geometry 43
  vibration 51
Monastery 5
  rule 5
Monosaccharide 113
Mouth feel 195, 200
Munich 7

NAD 170
Neolithic era 4
Neuron 198
  firing 200
Neurotransmitter 99
Neutron 33, 34
Nitrogen 237

Nucleation site   231, 235, 237
  Fusarium   237
Numbers   56
  exponential notation   56

Olfactory bulb   198, 200
Organic chemistry   89
Original gravity   178
Osmosis   81
  in cells   82
  osmotic stress   82
  reverse   81, 82
Oxidation   244, 252
Oxidation number   253
  balance   255
  determination   253
Oxide ion   255
Oxygen
  addition   27, 171
  air   260
  atmospheric   244
  exclusion   260
  permeation   261
  reactive species   255
  reduction   255, 257
  structure   255
  valence electrons   255
  and yeast   161
  yeast requirement   163

Package   13, 241
  bottle   243
  can   246
  cask   241
  keg   242
  pressure   241
Pasteur, Louis   9
Pasteurization   30, 249
Peptidase   145
Percent
  alcohol by volume   187
  weight   215
Periodic table   35, 40
  electronegativity   46
  group   35, 38
  main group   35
  period   35
  valence electrons   36

Peroxide radical   257
pH   10, 133, 218
  calculation   75
  charge   75
  glass electrode   191
  highest and lowest   73
  hydronium ion   73
  importance   73
  indicator   190
  ionization   75, 79
  measurement   73, 190
  meter   191
  neutral   75
  paper   190
  protein   73
  scale   75
Phenol   97
Photon   186
Phytic acid   261
Pietism   11
Pilsen   7, 218
Pipe scale   77
Points   178
Polarity
  bond   46
  cancellation   47
  molecular   47
  temporary   48
  water   47
Polyphenol   228
Polysaccharide   121
Polyvinylpolypyrrolidone (PVPP)   230
Process analyzer   188
Product   52
Prohibition   10, 11
Protein   145, 228
  coagulation   25
  denaturation   52, 145, 236
  enzyme   131, 135
  haze   145, 155
  head   145
  hydrolysis   145
  pH   73
  primary structure   137
  regions   137
  removal   145, 155
  transmembrane   195
  yeast   145

Proton 50
  atomic number 33
  charge 33
Psychoactive compounds 99
Pub 242
Pump 195
  ion 166
PVPP, see Polyvinylpolypyrrolidone
Pyruvic acid 168, 170, 172

R and S 117
Reactant 52
Reaction 36, 52
  cell 131
  temperature 52
Reactive oxygen species 255, 257
  metal ions 257
Receptor 196
  aroma 198
Recipe
  beer 290
  calculation 217
Redox 244, 252
  hydrogen 253
Reducing sugar 126
Reduction 244, 252
Refractive index 182
Refractometer 182
Reinsheitsgebot 7
Residual alkalinity 79
Resonance 167
  delocalization 167
  free radical 257
Respiration 161, 170, 171
  energy 171
  products 161
Rho-isohumulone 155
Rice 9, 144, 217

Saccharomyces 21
Saccharomyces cerevisiae 171
Saccharomyces pastorianus
  171
St. Columban 5
St. Gall monastery 5
Sanitize 280
Sanitizer 274, 286
  cost 276
  no-rinse 287

Schwann, Theodor 9
Sensory analysis 192
SI (International System) 57
Silica gel 230
  protein 230
Siphon 273
Skara Brae 4
Skunked 154. *See also* Lightstruck beer
Skunk-proofing 207
Sodium 38
Sodium carbonate 243
Sodium chloride 38
Sorensen, S. P. L. 10
Sparge/sparging 125, 150, 278
Specific gravity 177, 215
  alcohol effect 178
  beer style 178
  carbohydrate content 177
Squalene 162, 171
Stacking, PVPP 230
Stainless steel 241, 242, 244
  composition 244
Standard Research Method
  186
Starch 8, 113, 121, 126, 131, 149.
    See also names of specific
    starches
  hydrolysis 131, 132, 141
  iodine test 126
Stereoisomers 118
Sterol 162
Structure
  condensed 90, 258
  Lewis 89
  ring 91
  semicondensed 90
  skeletal 91
Sucrose 126
Sugar 8, 25, 101, 113, 149, 150, 171.
    See also names of specific
    sugars
  disaccharide 120
  from starch 132
  open-chain 114
  phosphorylation 168
  potential extract 217
Sulfite ion 261
Sumer 1, 3
Superoxide dismutase 257

Superoxide ion 256
  unpaired electron 256
Surface 225
  area 132
  decrease 227
  energy 225
  tension 227
Surfactant 227, 235
  amphiphilic 227
Synods of Aachen 5

Taste 195
  bitter 196, 197
  bud 196
  salty 196, 197
  sour 196, 197
  sweet 196, 197
  umami 196
Tasting 192
  analytical 192
  hedonistic 192
Tavern 13
Temperance 11
Temperature 183
  absolute 58
  conditioning 290
  control 28
  correction 271
  fermentation 6, 28, 265
  gelatinization 23
  kilning 20
  lager 289
  mashing 23, 25, 269, 277
  measurement 10
  molecular motion 51
  pitching 284
  reaction 52
  sparge 278
  staling 260
  strike water 277
  unit 58
Tetrahydro-isohumulone
  155
Thermometer 183–184
Threose 118
Threshold potential 200
Tollens's reagent 126
Trans-2-nonenal 252
  formation 257, 258

Transamination 172
Transition state 135
Traugott-Kutzing, Friedrich 9
Trub 157, 229

Umami 196
Underback 150
Units 57
  amount 60
  conversion tables 61
  derived 58
  energy 59
  force 59
  hardness 78
  metric 57
  mole 60
  pressure 59
  symbol 57
  volume 59

Valence
  electron 35
  shell 35, 37
Valency 40
  group number 40
  structure 41
Valentine's Day massacre 11
Vapor pressure 234
  temperature 234
Vegan(s) 230
Vicinal diketone 222
Viscosity 124
Vitalism doctrine 9
Volstead, Andrew 11
Volstead Act 11
Volume percent 65
Vorlauf 149, 150, 278

Water 17, 21, 36, 218, 274
  activated carbon 83
  adjustment 84
  alkaline 218
  alkalinity 77, 79
  as acid and base 72
  beer style 218
  boiling 84
  brewing 69
  cohesive 225
  demineralizing 82

Water (cont'd)
    disinfectant 79
    equilibrium 72
    filtration 80
    foam and haze 71
    formula 69
    hardness 77
    heating 71
    hydrogen bond 70
    ion exchange 82
    ion product 72
    iron removal 80
    molecule 69
    neutral 72
    oxygen removal 83
    phase 69
    polarity 70
    pollutant 69
    purification 80
    reverse osmosis 81
    shape 46, 69
    softening 82
    specific heat 71
    strike 277
    structure 41
    treatment 79
    vaporization 70
Weight-volume percent 64
Whirlpool 157
Widget 238

Wine 5
    ethanol content 10
Wood 121
Wort 125, 132, 149
    boiling 219
    chiller 158
    hopped 25
    separation 149, 150
    sweet 25

Yeast 124, 131, 161, 276, 280
    ale 276
    bottom fermenting 21
    Brettanomyces 21
    fermentation 9, 10
    lager 276, 290
    liquid 282, 283
    oxygen requirement 171
    pitch 27, 281
    POF+ 214
    respiration 170
    Saccharomyces 21
    S. cerevisiae 171
    S. pastorianus 171
    starter 284, 290
    strain 173
    strains 276
    sulfite ion 261
    top fermenting 21
    wild 27